SpringerBriefs in Molecular Science

Green Chemistry for Sustainability

Series Editor
Sanjay K. Sharma

T0214885

For further volumes:
http://www.springer.com/series/10045

SpringerBriefs in Molecular Science

Green Chemistry for Sustainability

Mohamed Ghoul · Latifa Chebil

Enzymatic Polymerization of Phenolic Compounds by Oxidoreductases

 Springer

Mohamed Ghoul
Laboratoire d'Ingénierie des Biomolécules
Nancy université, ENSAIA-INPL
2 Avenue de la Forêt de Haye
54505 Vandoeuvre les Nancy
France
e-mail: mohamed.ghoul@
          ensaia.inpl-nancy.fr

Latifa Chebil
Laboratoire d'Ingénierie des Biomolécules
Nancy université, ENSAIA-INPL
2 Avenue de la Forêt de Haye
54505 Vandoeuvre les Nancy
France
e-mail: latifa.chebil@ensaia.inpl-nancy.fr

ISSN 2191-5407
ISBN 978-94-007-3918-5
DOI 10.1007/978-94-007-3919-2
Springer Dordrecht Heidelberg New York London

e-ISSN 2191-5415
e-ISBN 978-94-007-3919-2

Library of Congress Control Number: 2012930373

Printed on acid-free paper

Springer is part of Springer Science+Business Media (www.springer.com)

Contents

Enzymatic Polymerization of Phenolic Compounds by Oxidoreductases

Abstract The enzymatic polymerization of phenolic compounds arouses more and more interest in the several fields such as food, cosmetic and pharmaceutical. The use of these compounds for their antioxidant properties is limited by their low solubility and thermal stability. The polymerization reaction improves their solubility and/or their thermal stability and provides new properties. These properties are dependent on the molecular mass and the structure of polymers. The reaction yield, the polydispersity, the molecular mass, the structure and thus the properties of synthesized polymers can be controlled by the mode of control of the reaction and by the reactional conditions. In this book, we analyze the key factors (temperature, solvent, origin of the enzyme, structure of the substrate, reactor design,...) who control the polymerization of phenolic compounds by the oxidoreductase enzymes, to obtain polymers with desired characteristics and properties.

Keywords Polymerization · Phenolic compounds · Oxidoreductase enzymes

1 Introduction

Polyphenols have received a great interest in food, pharmaceutical and cosmetic applications in the recent years. Due to their high antioxidant activity, it is assumed that they play a major role in the prevention of cardiovascular, cancer and degenerative diseases (Havsteen 2002; Heim et al. 2002). They also act as inhibitors of the activity of several enzymes and possess antibacterial properties (Kurisawa et al. 2003a). They are largely present in plant in the form of monomers or polymers structures. The main limitations for the use of polyphenols are their lower solubility and stability (Abou El Hassan et al. 2000; Friedman and Jurgens 2000; Zhu et al. 2002). Moreover, it has shown that, depending to their molecular

M. Ghoul and L. Chebil, *Enzymatic Polymerization of Phenolic Compounds by Oxidoreductases*, SpringerBriefs in Green Chemistry for Sustainability, DOI: 10.1007/978-94-007-3919-2_1, © The Author(s) 2012

weight and their structure, the biological activities of phenol oligomers are different. For these reasons, in vitro polymerization of phenols was the subject of several papers. They can be realized either by chemical way (Fulcrand et al. 1996; Kim et al. 2004a) or by enzymatic synthesis (Kobayashi and Higashimura 2003; Kobayashi et al. 2001). The enzymatic way is more attractive because it can be realized under mild operating conditions of pH and temperature and it exhibits a good regio-, chemo- and enantio-selectivity. Different classes of enzymes (oxidoreductases, transferases, hydrolases, lyases, isomerases or lipases) are used for the polymerization of phenolic species. In vitro synthesized oligomers are often characterized by better solubility (Bruno et al. 2005; Kurisawa et al. 2003c; Mita et al. 2002; Oguchi et al. 2000), thermostability (Ayyagari et al. 1995; Kurioka et al. 1994; Oguchi et al. 2000; Tonami et al. 2000) or antioxidant properties (Kurisawa et al. 2003a; Kurisawa et al. 2003c; Kurisawa et al. 2003e). The polymerization of phenolic compounds allows also to synthesis new materials such as new resin (Ikeda et al. 2000; Xia et al. 2003) or graft copolymers by copolymerization (Faure et al. 1995; Mai et al. 2000; Rittstieg et al. 2002). Kim et al. (2003b) showed that the polymerization of cardanol, using soybean peroxidase (SBP), produces films with excellent anti-biofouling activity against *Pseudomonas fluorescens* compared to other polymeric materials such as polypropylene. Furthermore, the complex oligo-catechin/sulfonated polystyrene (as template) shows inhibitory effect for the proliferation of colon cancer cell (Bruno et al. 2005). Suitable photoconductive material, as polyfluorophenol, was also synthesized by using horseradish peroxidase (HRP) biocatalyst (Zaragoza-Gasca et al. 2011).

The aim of this book is to synthesize the knowledge in the field of enzymatic polymerization of phenols using oxidoreductases. Both the processes used for the synthesis and the operating conditions will be presented and discussed. The performance of these processes will be summarized in synthetic way.

2 Mechanism of the Enzymatic Polymerization of Phenolic Compounds by Oxidoreductases

Oxidoreductive enzymes are able to transform phenols through oxidative coupling reactions with production of polymeric products by self-coupling or cross-coupling with other molecules. The two principal groups used in phenolic compounds polymerization process are the laccases (E.C 1.10.13.2) and the polyphenoloxidases (E.C. 1.14.18.1). The principal difference between these two groups is the nature of co-substrate used (Canfora et al. 2008).

2.1 Mechanism of Laccases Catalysis

Laccase can have different sources: fungi (*Trametes versicolor*, *Myceliophotore*, *Pycnoporus*,...) or in plant (*Rhus vernicifera*, *Pinus taeda*, ...) (Gianfreda et al.

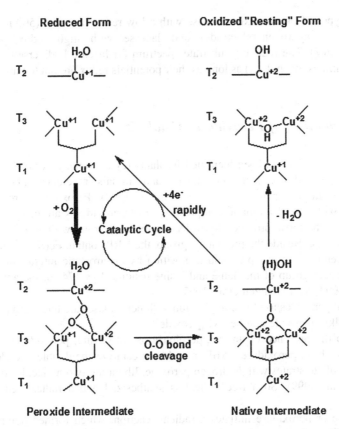

Fig. 1 Catalytic cycle of laccase showing the mechanism of four-electron reduction of a dioxygen molecule to water at the enzyme copper sites. T: type. Adapted from (Shleev et al.) with kind permission of © The American Chemical Society (2006)

1999; Solomon et al. 1996; Uzan et al. 2011). Laccase can catalyzes the oxidation of a variety of compounds, including ortho- and para-benzenediols, polyphenols, aminophenols, polyamines, lignin, aryldiamines and number of inorganic ions (Burton 2003; Claus 2004; Morozova et al. 2007; Riva 2006). Laccase are multicopper oxidase, they contain at least one blue copper or T1 site and a T2/T3 trinuclear cluster as the minimal functional unit. This enzyme couple the four-electron reduction of dioxygen to water with the oxidation of substrate. The substrate is oxidized by one electron, which creates a radical species. Laccase initializes radical reaction outcome with polymer formation. Figure 1 shows the catalytic mechanism of laccase involving a four- electron reduction of the dioxygen molecule to water at the enzyme copper sites (Shleev et al. 2006; Solomon and Yoon 2008).

The oxidation can be controlled by redox potential differences between the reducing substrate and the T1 Cu in laccase (Xu 1996).

Two types of laccase exist: laccase with a low redox potential (500 mV compared to a hydrogen electrode) and laccase with high redox potential (700–800 mV). The reducing substrate spectrum for laccase is diverse (ferrocyanidine, anilines, phenols…) as long as their potentials redox are not too all (>1 V).

2.2 Mechanism of Peroxidases Catalysis

Peroxidase activity has been identified in plants (cytochrome c peroxidase, plant ascorbate peroxidase,…), microorganisms and animals (myeloperoxidase, lactoperoxidase, thyroid peroxidase, …) (Welinder 1992). Peroxidases are able to catalyze oxidation of aromatic compounds. Most peroxidases are heme enzymes and contain the ferric protoporphyrin IX (protoheme) group with an iron atom in their active site. Beside the protoheme group, the HRP contain Fe(III) atom, in the reactive center. The group is held in position by electrostatic interaction of one propionic acid group of the heme and lysine residue (Lys 174) of the apoprotein (Poulos 1993; Van Deurzen et al. 1997).

Typical peroxides used in combination with peroxidases are hydrogen peroxide (H_2O_2), alkyl peroxide and benzyl peroxide.

Due to the presence of an iron-porphyrin as active site, peroxidases such as HRP or Soybean peroxidases (SBP) are able to catalyze the double one-electron-oxidation of substrates with hydrogen peroxide, liberating two molecules of water (Lalot et al. 1999). So a free radical is synthesized and a radical reaction is initiated.

Laccase or peroxidase initiates a radical reaction, which formed polyphenols species. The radical transfer is not precisely described, but the reaction's conditions could be controlled by the polydispersity, the molecular weight of polymers synthesized and the type of linkages between the monomers (Kobayashi and Higashimura 2003).

2.3 Mechanism of Phenols Polymerization

As previously described, the function of the enzyme is to produce the phenoxy radicals under mild reaction. The resulting phenoxy radicals are able to form polymers via recombination process. The oxidative polymerization of phenols is basically a polycondensation reaction. A model for the polymerization was developed, which divides the polymerization into four steps (Fig. 2) (Dec and Bollag 1990; Job and Dunford 1976; Nayak 1998; Reihmann and Ritter 2001).

The first step is the formation of phenoxy radicals by the HRP-catalyzed oxidation of phenols (Fig. 2a). It is the only step that is controlled by the enzyme kinetics. The phenoxy radicals form dimers by recombination (Fig. 2b). At the beginning of the reaction, practically all phenols are converted to dimers.

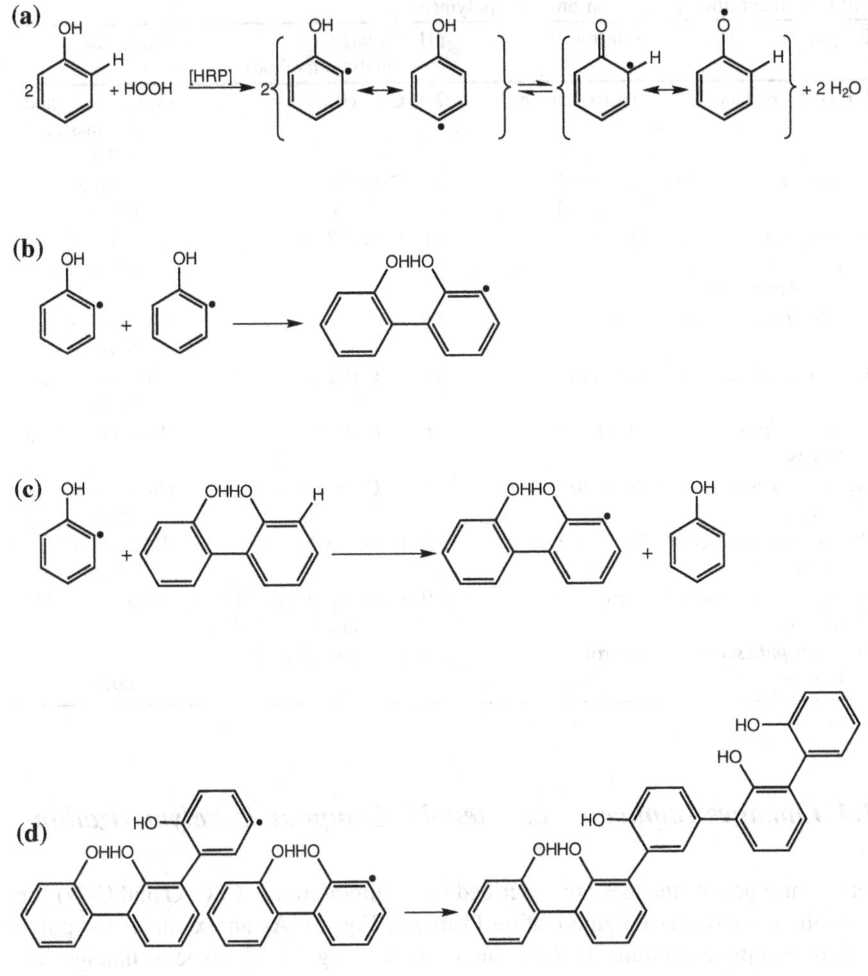

Fig. 2 Mechanism of phenol polymer formation, HRP: Horseradish peroxidase. **a** Formation of free phenoxyl radicals. **b** Recombination of the phenoxyl radicals. **c** Radical transfer. **d** Chain grow by alternating radical transfer and recombination (repetition of step 2)

When the concentration of free phenoxy radicals is decreasing, an electron-transfer reaction is more likely than further recombination. This leads to the formation of oligomer radicals (Fig. 2c), which then form oligomers of even higher molecular weight by recombination (Fig. 2d). The radical transfer reaction of a phenoxy radical and an oligomer regenerates a phenol monomer, which can be oxidized again by the enzyme to initiate new radical transfer reactions. When the phenoxy radical is not reacting fast enough in a recombination or a radical transfer step, oxidation may take place leading to ketone structures.

Table 1 Indentified linkages in phenolic polymers

Enzyme	Substrate	pH	Linkage (analytic method)	Reference
R.vernificera laccase	Hydroquinone	7.4	C–C (RMN)	(Witayakran and Ragauskas 2009)
Picnoporus coccineous laccase	2,6-dimethyl-phenols	5.0	C–O (IR)	(Ikeda et al. 1996a)
Horseradish peroxydase, Soybean peroxydase	Quercetin	7.0	C–O (FTIR)	(Fenoll et al. 2003)
Myceliophtora laccase	Rutin	5.0	–	(Kurisawa et al. 2003c)
Myceliophtora laccase	Catechin	5.0	C–C (FTIR)	(Kurisawa et al. 2003e)
Trametes villosa laccase	Bisphenol A	6.0	C–C (RMN)	(Mita et al. 2003)
Trametes versicolor laccase	α-naphtol	5.0	C–O (FTIR)	(Aktas et al. 2000)
Picnoporus coccineous laccase	Syringic acid	5.0	C–O (RMN)	(Ikeda et al. 1996b)
Picnoporus coccineous laccase	Phenols	5.0	C–O suggested (FTIR) C–O and C–C (FTIR)	(Mita et al. 2003)
Trametes pubescens laccase	Totarol	5.0	C–O and C–C (RMN)	(Ncanana et al. 2007)

2.4 Linkages Implicated in Phenolic Compounds Polymerization

Different types of linkages are described in the literature (C–C, C–O and C–N) for phenolic compounds polymerization (Table 1, Fig. 3). As an example, for poly-flavans, simple and double linkages are observed (Fig. 3). The type of linkage can be changed depending on synthesis conditions. Acid pH will promotes C–C linkages, while C–O linkages are favored at basic pH (Kobayashi and Higashimura 2003). The nature of solvent affects also both the position (Intra et al. 2005) and the type of the linkage (Mita et al. 2003; Ncanana et al. 2007). This aspect of linkage control will be discussed in paragraph 4.

3 Different Systems of Polymerization

Four systems of enzymatic polymerization of phenolic compounds were identified. Monophasic system, biphasic system, reversed micelle system and Langmuir–Blodgett system.

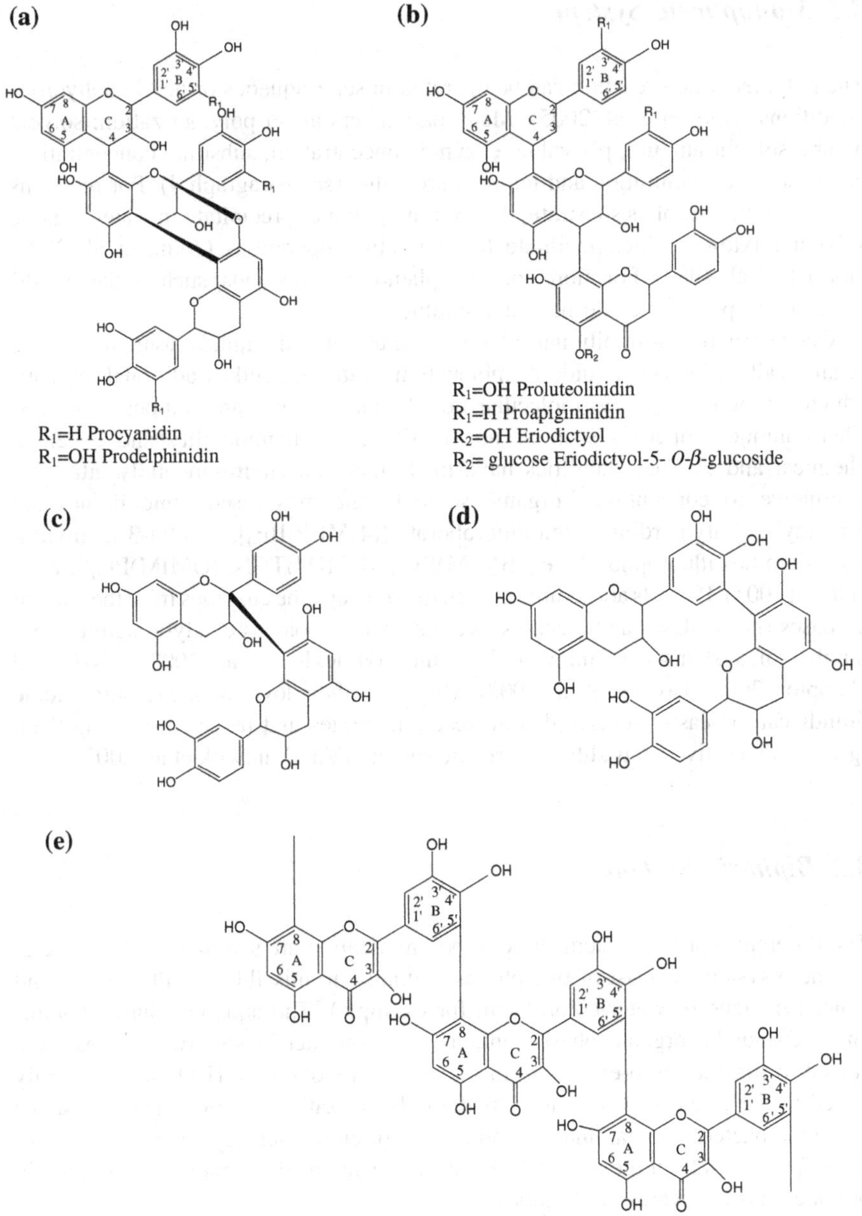

(a)

R$_1$=H Procyanidin
R$_1$=OH Prodelphinidin

(b)

R$_1$=OH Proluteolinidin
R$_1$=H Proapigininidin
R$_2$=OH Eriodictyol
R$_2$= glucose Eriodictyol-5- *O-β*-glucoside

(c)

(d)

(e)

Fig. 3 (**a**) Heteropolyflavan-3-ols from Ruby Red sorghum [*Sorghum bicolor* (L.) Moench].
(**b**) Glucosylated heteropolyflavans with a flavanone, eriodictyol or eriodictyol-5-*O-β*-glucoside,
as the terminal unit from Ruby Red sorghum [*Sorghum bicolor* (L.) Moench], (**c** and **d**) Di-
catechins, (**e**) polyquercetin

3.1 Monophasic System

The polymerization reaction can be operated in semi-aqueous or nearly anhydrous conditions (Hudson et al. 2005). Many parameters affect polymerization: solvent nature, solvent amount, pH value, enzyme concentration, substrate concentration, co-substrate concentration and temperature value (see paragraph 4). For reactions realized with phenol as substrate, the forming polymer precipitate in water/organic solvent mixtures, which facilitate the extraction operations (Akita et al. 2001; Dordick et al. 1987). For more complex phenolic compounds, such as flavonoids, the forming polymer could be water soluble.

One alternative of traditional solvent is the use of ionic liquids. Ionic liquids are organic salts, which are liquids at ambient temperatures. Unlike traditional solvents, which can be described as molecular liquids, ionic liquids are composed of ions. Their unique properties such as no volatility, non-flammability and excellent chemical and thermal stabilities have made them an environmentally attractive alternative to conventional organic solvent. The most used ionic liquids are 4-methyl-N-butylpyrdinium tetrafluoroborate [(4-MBP)BF$_4$], 1-butyl-3-methylimdizaolium hexafluorophosphate [(BMIM)PF$_6$], (BMIM)Tf$_2$N, (OMIM)PF$_6$) (Yang and Pan 2005). Phenols are soluble in such medium and the enzymes from the class of oxidases (peroxidases and laccases) were shown to possess catalytic activity in a number of systems containing ionic liquids (Hinckley et al. 2002; Laszlo and Compton 2002; Tavares et al. 2008). Owing to their low vapor pressures, ionic liquids can be easily separated from reaction solutes and recycled, making them 'green" alternatives to traditional organic solvent (Van Rantwijk et al. 2003).

3.2 Biphasic System

The different biphasic systems used in polymerization are summarized in Table 2.

These systems consist of two phases mutually no miscible (HEPES buffer and water-isooctane or water-chloroform, for example). The aqueous phase contains enzyme and the organic phase contains the monomer. These two phases were stirred to form a micro-emulsion. In the case of peroxidises, H$_2$O$_2$ was gradually added to the micro-emulsion to start the polymerization reaction. These systems avoid the bacterial contamination and the use of emulsifier (Ayyagari et al. 1996). These processes limit the inhibition by substrate and the enzymatic denaturation in organic phase (Tavares et al. 2008).

In the case of laccase, it reported that the use of biphasic system (acetate buffer with non-miscible solvents AcOEt, CHCl$_3$, methyl t-butyl ether, t-amyl alcohol or toluene) affects the performances of the oxidation reaction (Ncanana et al. 2007; Tavares et al. 2008). Low rates of oxidation in a biphasic system can be attributed to insufficient mass transfer of the reactants between the two phases (Carrea 1984).

3.2.1 Reversed Micelle System

Nanostructured reversed micelles induce a high enzyme activity in organic solvents. In this system, enzymes maintain their highly tridimensional structure and their activity are 20-fold high than in water (Martinek et al. 1982). The most used system is composed by isooctane/water/sodium dioctylsulfosuccinate (anionic surfactant = AOT) (Fig. 4) (Ayyagari et al. 1995, 1996; Dubey et al. 1998; Premachandran et al. 1996). In this systems, surfactants such as poly(ethyleneimine) bromide and cetyl bromide are used (Khmelnitsky et al. 1992).

The monomer conversion and polymer yields are better or equal, with reversed micelle system, to those obtained in dioxane/water medium (Table 3). The molar ratio of water to surfactant and the presence of salts affected the monomer conversion yield in micelle system. The presence of LiBr (at ≥ 0.35 w/v %) decreases the molecular weight average (Mw) of polymers (Ayyagari et al. 1995).

3.2.2 Langmuir–Blodgett

Langmuir–Blodgett is an operation that consists to orientate amphiphilic molecules to water–air interface in order to obtain a structured monolayer. Such organization allows controlling link's type between monomers, and preventing the formation of para–para structures (Akkara et al. 1992; Bruno et al. 1995; Reihmann and Ritter 2000). The synthesized polymers have better thermostability and electronic properties than monomers (Bruno et al. 1995).

4 Effects of Operating Conditions on Enzymatic Polymerization of Phenolic Compounds

The activity of the enzyme and the solubility of the substrate are the two important factors for high polymerization reactions performances. The natural environment of the enzyme is water while the phenolic substrates are more soluble in organic solvents. It is therefore necessary to seek a compromise between these two criteria. In this section, we analyzed and summarized the effects of several operating parameters on the performances and the regioselectivity of the polymerization (Table 4, non-exhaustive list).

4.1 Effect of Enzyme

4.1.1 Origin and Form of Enzyme

The type (laccase, peroxidase) and the origin (HRP, SBP, CCP,...) of the oxidoreductase used during the polymerization can affect the performance of this reaction, the structure and the properties of the obtained polymers (Table 5). Thus,

Table 2 Different systems of enzymatic polymerization of phenolic compounds

System	Solvent/phase	Enzyme	Substrate	Observations	References
Monophasic system	1,4dioxane, DMF or methanol with water	HRP	phenol	Higher yield: 1,4 dioxane 20–60%, DMF 20 or 40% at 30°C, 4 h	(Akita et al. 2001)
	Dioxane/acetate buffer	HRP	phenol	Activity of enzyme up to 95% of solvent	(Dordick et al. 1987)
	Tert-butanol, Ionic liquid: [(4-MBP)BF$_4$], [(BMIM)PF$_6$]	Laccase C, HRP, SBP	Syringaldazine, 2-methoxyphenol, anthracene	Higher yield of anthracene oxidation by laccase C in [(4-MBP)BF$_4$] compared at Tert-butanol medium, in presence of mediator	(Hinckley et al. 2002)
	Ionic liquid: [bmim][Tf$_2$N], [bmim][PF$_6$], [omim][PF$_6$]	Microperoxidase-11	2-methoxyphenol	Higher activity in [bmim][Tf$_2$N]	(Laszlo and Compton 2002)
	Succinate buffer/1,4 dioxane	Glucose oxidase/HRP	phenol	Higher yield: 1,4 dioxane/succinate buffer (15/10)	(Uyama et al. 1997)
	Phosphate buffer/DMF (80/20%)	Tri-enzymatic system: invertase/glucose oxidase/SBP	p-cresol, sucrose	The three enzymes are immobilized on glass microfluidic channels, the kinetic properties are nearly identical to SBP catalysis in solution	(Lee et al. 2003)
Biphasic system	Isooctane/buffer	HRP	p-ethylphenol	Good monomer conversion, fair polymer yield	(Ayyagari et al. 1996)
	Sodium acetate buffer/solvent (v:v) Solvent: AcOEt, CHCl$_3$, methyl tert-butyl ether, t-amyl alcohol or toluene	Laccase Trametes pubescens	totarol	The reaction were observed to proceed very slowly	(Ncanana et al. 2007)

(continued)

Table 2 (continued)

System	Solvent/phase	Enzyme	Substrate	Observations	References
Reversed micelle	AOT/isooctane AOT/isooctane/CHCl$_3$	HRP	p-ethylphenol	Higher yield: 100% isooctane Higher Mw: isooctane/CHCl$_3$ (25/75%)	(Ayyagari et al. 1995)
	AOT/isooctane	laccase	o-chlorophenol	Laccase hosted in the AOT reversed micelle exhibited high catalytic activity in isooctane, whereas lyophilized laccase did not exhibit catalytic activity in organic solvents.	(Michizoe and Goto 2001)
Langmuir–Blodgett	HEPES buffer	HRP	Aniline, phenol	The polymers are ordered and have a good thermal stability and significant optical and electronic properties.	(Bruno et al. 1995)

DMF dimethylformamide, *HRP* Horseradish peroxidase, *SBP* Soybean peroxidase, *AOT* dioctylsodiumsulfosuccinate, [bmim][Tf$_2$N]: 1butyl-3-methylimidazolium bis(trifluoromethylsulfonyl)imide, [bmim][PF$_6$]: hexafluorophophates of 1-octyl-3methylimidazolium, [omim][PF$_6$]: hexafluorophophates of 1-octyl-3methylimidazolium, [(4-MBP)BF$_4$] : 4-methyl-N-butylpyridiniul tetrafluoroborate, [(BMIM)PF$_6$] : 1-butyl-3-methylimdizaolium, *HEPES* 4-(2-Hydroxyethyl)piperazine-1-ethanesulfonic acid

No structuration

Langmuir Blodgett film balance

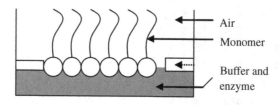

Reversed micelle

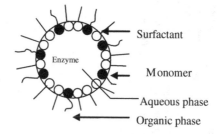

Association by Layer -by-Layer

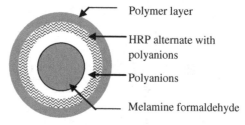

Template association

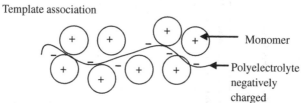

Fig. 4 Types of association used in enzymatic polymerization

Table 3 Effect of solvent composition on polymer molecular weight and polymer yield (%)

Substrate	Enzyme	Synthesis medium	Mw	Monomer conversion (%)	Polymer yield (%)	References
p-ethylphenol	HRP	85/15 dioxane/water	3,000	80	15	(Ayyagari et al. 1995)
		AOT reversed micelle in 100% isooctane (W_0:15)	2,500	100	100	
		AOT reversed micelle in 100% isooctane (W_0:9)	2,500	90	100	
		AOT reversed micelle in 100% $CHCl_3$ (W_0:7; phase separation)	1,000	20	10	
p-ethylphenol	HRP	Isooctane/buffer biphasic system	1,700			(Ayyagari et al. 1996)
Catechol	HRP	80/20 dioxane/phosphate buffer			18	(Dubey et al. 1998)
		AOT reversed micelle in 100% isooctane			36	
Ethylphenol	HRP	85/15 dioxane/water		95	~20	(Rao et al. 1993)
		AOT reversed micelle in isooctane (W_0:15)		~95	~95	
Ferulic acid	MTL	Ethyl acetate/sodium phosphate buffer (80:20)		84		(Mustafa et al. 2005)
Totarol	TPL	Sodium acetate buffer/solvent (v/v) Solvent: AcOEt, CHCl3, methyl tert-butyl ether, t-amyl alcohol or toluene		<10		(Ncanana et al. 2007)

AOT bis3-ethylhexyl sulfosuccinate sodium salt, W_0 molar ratio of water to surfactant, Mw weight-average molecular mass, MTL laccase of Myceliophthora thermophila, TPL laccase of Trametes pubescens, HRP Horseradish peroxidise, a: monomer converted/monomer added initially, b: ratio of the amount of polymer recovered as an insoluble fraction in isooctane to the amount of monomer converted

Table 4 Overview of factors affecting the enzymatic polymerization of phenols (non-exhaustive list)

Catalyst	Substrate	Factors					Note	References
		Solvents (%)	Buffer (pH)	Cofactor	Temperature (°C)	Time		
TBL (3 U/ml)	Totarol (0.5 g/l)	Methanol, acetonitrile, acetone, AcOEt, CHCl$_3$, methyl tert-butyl ether, t-amyl alcohol, toluene (50)	Δsodium acetate buffer (4–7)	Under air	20-30-40-50	24 h	Optimal: pH 4.5-5, 30°C, acetone	(Ncanana et al. 2007)
TVL (0–275U/ L)	Δpyrogallol (0–3 g/l)		ΔTampon sodium acetate (3–7)	ΔO$_2$ (0–20 mg/l)	Δ 25–55	60 min	Optimal : pH 4.5, 45°C and 18 mg/l dissolved oxygen	(Guresir et al. 2005)
TVL	Δα-naphtol (23–13636 g.m^{-3})	Δacetone (0–80)	Δsodium acetate buffer (3–5) sodium phosphate buffer (6–8)	ΔO$_2$ (12–20 g.m^{-3})	24–50	Up to stabilization	Optimal: 50%acetone, pH 5, 3409 gm^{-3} monomer, 20.3 gm^{-3} in dissolved O$_2$, 0.173 U/cm^{-3} Average molecular weight 4920 Da	(Aktas et al. 2000)
TVL (25U/L)	Catechol (0.3 g/l)	Δacetone (10–20)	ΔTampon sodium acetate (4.5–5.5)	O$_2$	Δ 35–45	60 min	Optimum conditions estimate by RSM at 43.5°C, pH 4.87, 13.5% acetone, 300 mg/ml catechol and 0.025 U/ml to obtain 0.128 mg DO/min L for initial oxidation rate	(Aktas 2005)

(continued)

Table 4 (continued)

Catalyst	Substrate	Factors			Temperature (°C)	Time	Note	References
		Solvents (%)	Buffer (pH)	Cofactor				
TVL (0–30 U/L)	Δcatechol (0.025–10 g/l)	Δacetone (0–70)	Δsodium acetate buffer (3–5) sodium phosphate buffer (6–8)	ΔO_2 (250; 10 mg/l)	Δ 25–60	250 min	Maximal initial rate : 10% acetone, pH 5, 25°C, 0.02 U/ml, 250 mg/L S, 10 mg/l dissolved oxygen	(Aktas and Tanyolaç 2003)
ΔMYL ($1.45\ 10^{-3}$; $5.5\ 10^{-3}$; $14.5\ 10^{-3}$ U/L)	Rutin (10 g/l) derivate of rutin	ΔMethanol (30;50;70;100)	Δ phosphate buffer(7) acetate buffer (5)	Under air	Room temperature	24 h	Higher yield (81%) with pH 5, 30% MeOH, $5.5\ 10^{-3}$ U/L. Higher Mn ($11\ 10^{-3}$) with pH 5, 100% buffer, $5.5\ 10^{-3}$ U/L	(Kurisawa et al. 2003c)
ΔMYL ($1.2\ 10^3$;$2.0\ 10^3$ $2.8\ 10^3$;$3.6\ 10^3$ U/L)	Catechin (g/l)	ΔAcetone (3–20) Ethanol (5) ΔMethanol (3–20) Isopropanol (5)	Acetate buffer (5)	Under air	Room temperature	24 h	Higher yield: 95% (80% acetate buffer pH5, 20% acetone) Higher Mn: Mn:7000–8000 (93% acetate buffer pH 5, 7% acetone)	(Kurisawa et al. 2003e)

(continued)

Table 4 (continued)

Catalyst	Substrate	Solvents (%)	Buffer (pH)	Cofactor	Temperature (°C)	Time	Note	References
PCL (2.95 mg)	Syringic acid (20 g/l)	Acetone 50 Acetonitril 50 1,4-dioxane50 ethanol 50 methanol 50 methyl ethyl ketone 50 THF 50 Acetone 40 acetone/ chloroform/ buffer 33/17/50 acetone/ chloroform/ buffer 25/37.5/ 37.5	Acetate buffer (5)	Under air	Room temperature	24 h	Higher yield: 86% (50% 1,4-dioxane, 50% acetate buffer pH5) Higher Mw: 18000 (25% acetone, 37.5% chloroform, 37.5% acetate buffer pH5)	(Ikeda et al. 1996b)
PCL 50 μl	Phenol m-cresol bisphenol A 4-TBP (30 g/l)	Methanol 2-propanol 1-propanol ethylene glycol 1,4-dioxane acetone 50	Acetate buffer (5) in an equivolume mixture solvent and buffer		Room temperature	24 h	Higher conversion with 4-TBP in acetone/buffer (50:50): 100% Higher Mn with bisphenol A in 2-propanol/ buffer (Mn: 21300)	(Mita et al. 2003)
PCL Suberase 300 nkat/g subtrate)	Rutin (3 g/l)	glycerol/ethanol/ buffer in the ratio 1:1:2 (w/w/w)		Under air	30	24 h	Yield with PCL: 67% Yield with Suberase: 97%	(Uzan et al. 2011)

(continued)

Table 4 (continued)

Catalyst	Substrate	Factors					Note	References
		Solvents (%)	Buffer (pH)	Cofactor	Temperature (°C)	Time		
TVL, THL (30 nkat ml^{-1} laccase activity)	Calcium lignosulfonate (20 g/l)	Double distilled water	–	HBT (0.5 mM)	30	17 h	The Mw increase by 74% with THL and by 370% with TVL	(Nugroho Prasetyo et al. 2010)
Urushi laccase (10 mg)	Lignosulfonate (50 mg)	Ethanol (70)	Phosphate buffer (0.1 M; pH 7)	Under air	30	48 h	Yield: 86%	(Yoshida et al. 2009)
MTL (0.6 U/ml)	Calcium lignosulfonates (2 g/l)	–	Phosphate buffer (pH 5)		50	6 days	Molecular weight increase: 24%	(Kim et al. 2009)
ΔRVL (12.5 µg/ml)	(+) Catechin (30 mg)	Isooctane (95) Hexane (95) Toluene (95) Dichloromethane (95)	Water	Under air	25	1 h	Higher yield in the presence of hexane: 0.91 10^{-2}%	(Ma et al. 2009)
	(-) Epicatechin (30 mg)						Higher yield in the presence of hexane 0.80 10^{-2}%	
	Catechol (30 mg)						Higher yield in the presence of dichloromethane: 0.34 10^{-2}%	
TVL (3 U/ml)	Rutin (3 g/l)	methanol (30) methanol (30)	-water -buffer (citrate or phosphate ammonium buffer 0.01 M at different pH (4, 5, 7)		10, 20, 50	24 h	Highest weight-average molecular mass, about 3900 g/mol was obtained for lowest pH and temperature set point.	(Anthoni et al. 2008)

(continued)

Table 4 (continued)

Catalyst	Substrate	Factors		Buffer (pH)	Cofactor	Temperature (°C)	Time	Note	References
		Solvents (%)							
TVL (3 U/ml)	Esculin (3 g/l)	Methanol (30)		water		20	72 h	weight-average molecular mass, about 1800 g/mol	(Anthoni et al. 2010)
PCL 2.95 mg HRP 10 mg, SBP	2,6-dimethylphenol (12.4 g/l)	ΔAcetone 20–90 1,4-dioxane 60 Ethanol 60 Methanol 60		Acetate buffer (3–12)	Laccase: under air Peroxidases:H_2O_2 is added to the mixture every 15 min for 10 min Under air	Room temperature	24 h	Laccase: Higher yield: 61% (40% 1,4-dioxane, 60% acetate buffer pH5) Higher Mw: 8140 (40% acetonitrile, 60% acetate buffer pH5) HRP: Higher yield: 33% (40%acetone, 60% acetate buffer) Higher Mw: 6290 (40% ethanol, 60% acetate buffer)	(Ikeda et al. 1996a)
Laccase of *Trametes* sp. (2U/ ml, 0.4 U/ml)	phenol (200 mmol/ L)	Δ Acetone (0–80)		McIlvaine buffer (pH 4.5)	Under air	30	24	In buffer: absence of polymerization In acetone/buffer: maximum yield obtained at the acetone content of 20–30%.	(Tanaka et al. 2010)

(continued)

Table 4 (continued)

Catalyst	Substrate	Factors					Note	References
		Solvents (%)	Buffer (pH)	Cofactor	Temperature (°C)	Time		
	m-methoxyphenol (200 mmol/L) p-methoxyphenol (200 mmol/L) o-methoxyphenol (200 mmol/L) 2,6 dimethoxyphenol (200 mmol/L)						In buffer: Highest initial rate with p-methoxyphenol In acetone/buffer: Maximum rate (about 35%) and yield (about 90%) with p-methoxyphenol in 20% and 50 acetone continent, respectively	(Ikeda et al. 1998)
PCL 2.95 mg, HRP, SBP 10 mg	4-hydroxybenzoic acid derivatives (16.8 g/l)	Acetone 40 Methanol 40 1,4-dioxane 40	Acetate buffer (3–12)	H_2O_2 (30%, 28 µl, 0.25 mM)is added to the mixture every 15 min for 10 min Under air	Room temperature	24 h	Laccase: Higher yield: 86% (50% 1,4 dioxane, 50% acetate buffer pH5) Higher Mw: 7700 (40%acetone, 60% acetate buffer pH5) HRP: Higher yield: 79% (40% acetone, 60% acetate buffer pH5) Higher Mw: 15100 (40% acetone, 60% acetate buffer pH7)	

(continued)

Table 4 (continued)

Catalyst	Substrate	Factors						Note	References
		Solvents (%)	Buffer (pH)	Cofactor	Temperature (°C)	Time			
HRP (10 mg)	Phenol (18.8 g/l)	Δ1,4-dioxane (60;80)	Acetate buffer (5) Succinate buffer (5.5) Phosphate buffer (7)	H_2O_2 formed in situ by glucose oxidase (10 mg): glucose 36.4 g/L. Under air or oxygen	Room temperature	48 h		The ratio of DMF-soluble part of polymers dependent on the polymerization condition, mainly the mixed ratio of 1,4-dioxane and a buffer. The higher Mn (14700) with pH 5, 80% 1,4-dioxane, in air atmosphere. Higher yield (78%) with pH 5.5, 60% 1,4-dioxane in oxygen	(Uyama et al. 1997)
		Acetonitril (15/10) Acetone (15/10) 1,4-dioxane (15/10) Ethanol (15/10) Ethyl acetate (15/10) Methanol (15/10) iso-propanol (15/10) 1,4 dioxane/ ethyl acetate (10/5/10; 7,5/7,5/10; 5/10/10)	Succinate buffer (5.5)					Higher yield 78% (60% methanol, 40% succinate buffer pH 5.5) Higher Mn 7500 (20% 1,4 dioxane, 40% ethyl acetate, 40% succinate buffer pH 5.5)	

(continued)

Table 4 (continued)

Catalyst	Substrate	Factors Solvents (%)	Buffer (pH)	Cofactor	Temperature (°C)	Time	Note	References
HRP (10 mg)	p-cresol p-ethylphenol p-n-propylphenol p-n-butylphenol p-n-pentylphenol at (18.8 g/l)	1,4 dioxane (60)	Succinate buffer (5.5)				Higher Mn (4400) with p-cresol Higher yield (72%) with p-n-pentylphenol	(Dordick et al. 1987)
HRP (95 10^3U/l)	Phenols p-methoxyphenol p-cresol p-chlorophenol 2,6-dimethylphenol 4,4'-biphenol aniline 1-naphtol 2-naphtol p-tert-butylphenol p-phenylphenol at 5–100 mM	ΔDioxane (0–100)	Acetate buffer (5)	H_2O_2, 20 μl at 30% (20 mM)			Activity of enzyme up to 95% solvent Average molecular weight up to $2.6\ 10^4$ Da with p-phenylphenol in 85% dioxane	
HRP (16 10^3 U/ L)	Phenols (2.235 g, 2.50 mmol)	Δ1,4-dioxane ΔDMF ΔMethanol 0, 20, 40, 60, 80, 100	distilled water	Solution at 30% H_2O_2 added 16 times at an interval of 10 min	10–60	10 min, 4 h, 48 h	Higher yield: >98% (1,4 dioxane 20–60%, DMF 20 or 40% at 30°C, 4 h)	(Akita et al. 2001)
HRP	4,4'-biphenyldiol	Acetonitrile Acetone 1,4-dioxane ethyl acetate isopropyl alcohol THF	Phosphate buffer (7)	H_2O_2 (batch)	Room temperature	24 h	Higher yield 97% (80% THF, 20% phosphate buffer pH7) Higher Mn 26000 (80% isopropyl alcohol, 20% phosphate buffer pH7)	(Kobayashi et al. 1996)

(continued)

Table 4 (continued)

Catalyst	Substrate	Factors Solvents (%)	Buffer (pH)	Cofactor	Temperature (°C)	Time	Note	References
HRP (10 mg)	Phenol (0.1 g)	–	Phosphate buffer (pH7, 0.1 M)	- H_2O_2 (5%, 3.4 ml, 5.6 mmol) was added dropwise for 2 h -Carbon nanotubes (MWNT, MWNT-OH and MWHNT-COOH) as template (0.1 g)	Room temperature	24 h	Higher yield (56%) and Mn 2300 in the presence of MWHT-OH	(Yun Peng et al. 2009)
HRP (0.1 g/l)	m-cresol (0.1 M:0.2 M)	ΔEthanol (0–100)	HEPES buffer	Dropwise addition H_2O_2 in excess of 30% compared to monomer	Room temperature	24 h	Higher yield (over 90%) and M_w (7000) with 20% ethanol	(Ayyagari et al. 1998)
HRP (1 mg)	Catechol (5 mg)	–	Potassium phosphate buffer (0.01 M, pH 4.3)	Diluted H_2O_2 (0.02 M) was added dropwise over a period of 1 h template (PSS) (5 mg)	Room temperature	–	Higher yield of 90%	(Nabid et al. 2010)
HRP, SBP 40–120 10^3 U/l	Catechol (1.32 g/ 1 l)	1,4-dioxane (80)	Phosphate buffer (7), acetate buffer (5)	30% H_2O_2(28 µl, 0.25 mM) addition every 15 min for 16 times	25°C	24 h	HRP < SBP Yield of polymer is pH-dependant Higher yield: 60.6% (phosphate buffer)	(Dubey et al. 1998)

(continued)

Table 4 (continued)

Catalyst	Substrate	Factors					Note	References
		Solvents (%)	Buffer (pH)	Cofactor	Temperature (°C)	Time		
HRP (2 mg)	Phenol (0.47 g, 5 mmol)	–	Phosphate buffer (0.1 M, pH 7)	H_2O_2 (5%, 3.4 ml, 5.6 mmol) was added dropwise for 2 h. Copolymers: Pluronic: L61, L62, L64 and F68.	Room temperature	1 h	Higher yield (96%) and Higher Mn (582 10^3) in the presence of Pluronic F68.	(Kim et al. 2008)
HRP (7.5 mg)	2-hydroxycarbazole (HC) (5 mmol)	Dioxane (70)	Phosphate buffer	H_2O_2	Room temperature	24 h	Mn (1850), Mw (2800)	(Bilici et al. 2010)
SBP (0.1 mg/ml)	p-cresol (20 mM)	1-butyl-3-methylimidazolium tetrafluoroborate (BMIM(BF$_4$)) (50) 1-butyl-3-methylpyridinium tetrafluoroborate (BMPy(BF$_4$)) (50)	Phosphate buffer (10 mM; pH 7)	H_2O_2 (20 mM) added dropwise over a period of 2.5 h	60	22.5 h	Yield of 80% for both ionic liquids. Higher Mn (1530) in the presence of (BMPy(BF$_4$)).	(Eker et al. 2009)
	p-Phenylphenol (20 mM)	(BMIM(BF$_4$)) (70) (BMPy(BF$_4$)) (10, 70)					Higher conversion yield (99.7%) in (BMPy(BF$_4$)) (10). Higher Mn (1860) in BMPy(BF$_4$)) (70)	

(continued)

Table 4 (continued)

Catalyst	Substrate	Factors Solvents (%)	Buffer (pH)	Cofactor	Temperature (°C)	Time	Note	References
HRP (17.6 U/l)	4-CHP, 4-TBP, 4-IPP, 4-EP, 4-MP (30 g/l)	Ethylene glycol 1,4dioxane 1,3-dioxane 2-propanol 1-propanol ethylene glycol DMF Methanol Acetone t-butyl alcohol ethylene glycol 50	Phosphate buffer (7)	H₂O₂ (3.2 ml 5%, 5.3 mmol) added dropwise over 2 h under air	Room temperature	3 h	With 4-TBP: Higher yield 98% (50% 1,4-dioxane, 50% phosphate buffer pH7 Higher Mw 2400 (50% 1-propanol or t-butyl alcohol, 50% phosphate buffer pH7) Higher Mw 2400 (50% 1-propanol or t-butyl alcohol, 50% phosphate buffer pH7) After 1 h the precipitate materials were collected	(Mita et al. 2004)
ΔHRP 10–50 mg (10U/mg)	Phenols 10.6 mmol	ΔMethanol 25,50,75	4, 5, 8 (phosphate buffer for pH 7)	ΔH₂O₂ added dropwise for 25 h	Room temperature	30 h	Enzyme origin, buffer pH, mixed ratio of alcohol and buffer, purity and amount of HRP, concentration and addition rate of hydrogen peroxide strongly affected the molecular weight and solubility of the polymer. Higher yield (80%) in 2-propanol/buffer (pH5) (1:1), and higher Mw (8 10⁻³) in 2-propanol/buffer (pH 7) (1:1)	(Oguchi et al. 2000)

(continued)

Table 4 (continued)

Catalyst	Substrate	Factors			Temperature (°C)	Time	Note	References
		Solvents (%)	Buffer (pH)	Cofactor				
HRP (20 mg/ 50 ml)	Phenols with different substituents (32.8 g/l): p-CH$_3$ p-n-C$_3$H$_7$ p-i-C$_3$H$_7$ p-t-C$_4$H$_9$ p-n-C$_5$H$_{11}$ p-n-C$_7$H$_{15}$ o-i-C$_3$H$_7$ m-i-C$_3$H$_7$ H	1,4-dioxane (80)	Phosphate buffer (7)	0.5 mmol (0.56 µl) of hydrogen peroxide was added 20 times after 15 min intervals	Room temperature	24 h	Higher yield with p-i-C$_3$H$_7$ (95%) Higher Mn with no substituent phenol 35000	(Kurioka et al. 1994)
HRP	m-ethynylphenol 24 g/l	Methanol (50)	Phosphate buffer (7)	Hydrogen peroxide (5%,1.7 ml, 2.5 mmol) added dropwise for 2 h	Room temperature under air	3 h	Mw: 1700	(Tonami et al. 2000)
HRP SBP 20 mg in 25 ml	Cardanol 24 g/l	Methanol Ethanol 2-propanol t-butanol 1,4-dioxane 50	Phosphate buffer (7)	H$_2$O$_2$ added continuously by a perfusion pump for 6 h	Room temperature	24 h	With SBP: Higher yield 62% (50% 2-propanol, 50% phosphate buffer pH7) Higher Mw 12808 (50% methanol, 50% phosphate buffer pH7) HRP is inefficient for polymerization of caradanol	(Kim et al. 2003b)

(continued)

Table 4 (continued)

Catalyst	Substrate	Factors		Buffer (pH)	Cofactor	Temperature (°C)	Time	Note	References
		Solvents (%)							
HRP 8.8 U/ml SBP 8.32 U/ml	Bisphenol Dihydroxydiphenylmethane 5.0 mmol in 25 ml	Acetone 2-propanol methanol acetone 50		Phosphate buffer (7)	H_2O_2 (5%, 3.5 ml, 5 mmol) added dropwise to the reaction mixture for 2 h	Room temperature	3 h	With bisphenol HRP Higher yield 100% (50% 2-propanol, 50% phosphate buffer pH7) Higher Mw 5400 (50% 2-propanol, 50% phosphate buffer pH7) SBP Higher yield 99% (50% 2-propanol, 50% phosphate buffer pH7) Higher Mw 3040 (50% 2-propanol, 50% phosphate buffer pH7)	(Uyama et al. 2002)

Δ: parameter variation (15/10): ratio 15 solvent and 10 buffer, *Mw* weight-average molecular mass, *Mn* number-average molecular mass, *DMF* dimethylformamide, *THF* tetrahydrofuran, *HRP* Horseradish peroxidase, *SBP* Soybean peroxidase, *TVL* laccase *Trametes Versicolor*, *PCL* laccase *Picnoporus coccineous*, *CCP* laccase *Coprinus cinerius*, *TBL* laccase *Trametes pubescens*, *MYL* laccase *Myceliophtora*, *THL* laccase *Trametes hirsute*, *RVL* laccase *Rhus verniciferT*

Table 5 Polymerization of 2, 6-dimethylphenol or 4-hydroxybenzoic acid by different sources of catalyst in acetone and acetate buffer (40:60% vol.), at room temperature and 24 h under air

Monomer	Enzyme	Species	Yield (%)	Mw	PD	Reference
2,6-dimethylphenol	Peroxidase	HRP	33	57,600	1.8	
		SBP	38	81,000	1.8	(Ikeda et al. 1996a)
	Laccase	CCP	57	5,400	2.0	
4-hydroxybenzoic acid	Peroxidase	HRP	79	13,200	–	
		SBP	72	14,700	–	(Ikeda et al. 1998)
	Laccase	CCP	62	11,500	–	
		PCL	80	7,700	–	
		MYL	83	9,600	–	
		POL	O	–	–	

Mw weight-average molecular mass, *PD* polydispersity, *CCP* laccase *Coprinus cinerius*
PCL laccase *Pycnoporus coccineus*, *MYL* laccase *Myceliophthore*, *POL* laccase *Pyricularia oryzae*, *HRP* Horseradish peroxidase, *SBP* Soybean peroxidase

Ikeda et al. (Ikeda et al. 1998, 1996a) observed, in the presence of 2,6-dimethylphenol or 4-hydroxybenzoic acid as substrates, different yields of conversion and Mw, depending on the type of the used catalyst (laccase or peroxidase). Moreover, these authors noted that the Mw of the synthesized polymers is affected by the origin of the peroxidase (HRP or SBP). The origin of the enzyme also affects the conductivity of the synthesized polymers. Thus, the conductivity of poly (catechol), is 5.00×10^{-9} and 0.25×10^{-9} S.cm^{-1} when SBP and HRP were used, respectively (Dubey et al. 1998). For a given enzyme, the activity also varies depending on whether it is used in free form, immobilized, co-lyophilized or in the presence of surfactants.

4.1.2 Free Form

For polymerization reactions in organic solvents, the use of enzymes in free form is possible by addition of a surfactant, or by its co-lyophilization with a hydrophobic molecule. Indeed, these procedures can limit enzyme denaturation. Thus, Paradkar and Dordick (1994a, b), Kamiya et al. (2000), Bindhu and Emilia Abraham (2003) have shown that the activity of oxidoreductases is maintained, even in an environment almost dry, thanks to their combination with an ionic surfactant molecule. To avoid the toxic effect of H_2O_2, Angerer et al. (2005) use *tert*-butyl hydroperoxide, which is more soluble in the reaction medium. Under these conditions, a total turnover of 5000 µmol of substrate converted/µmol HRP was obtained without any change in final polymer molecular weight compared to reactions carried out in reverse micelles.

The activity of oxidoreductases also varies depending on whether they are predissolved in an aqueous solution or in solid form, dispersed directly into the reaction medium (Kamiya et al. 2000; Paradkar and Dordick 1994b). Lyophilization of the enzyme in the presence of a substrate, or a hydrophobic compound,

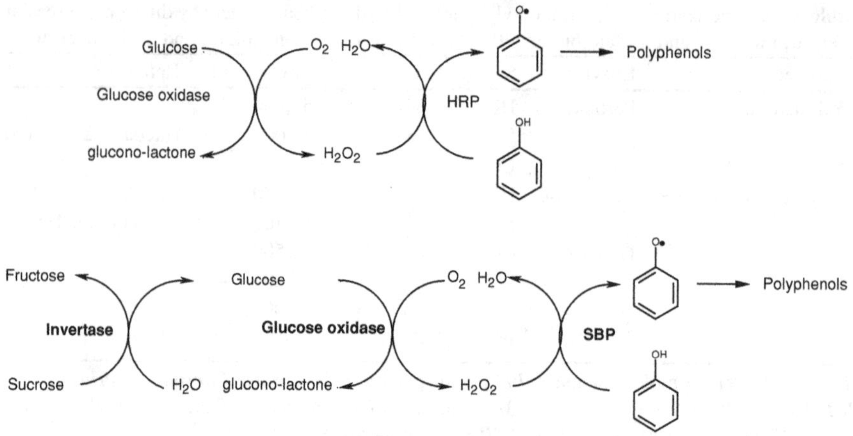

Fig. 5 Polymerization of phenol by bi-enzymatic (Uyama et al. 1997) and tri-enzymatic system (Lee et al. 2003) (HRP: Horseradish peroxidase; SBP: Soybean peroxidase). Reprinted with kind permission of © Nature Publishing Group (1997)

or a hydrophilic lyoprotectant increases its activity. Co-freeze increases the regioselectivity. Yu and Klibanov (2006) observed greater regioselectivity of HRP lyophilized at pH 9.5, in the presence of 75 mM D-proline, compared to the same enzyme non co-lyophilized.

4.1.3 Immobilized Form

To reduce the denaturant effect of organic solvents, different types of material were used to immobilize the oxidoreductases (Sepharose (Milstein et al. 1989), chitosan (Bindhu and Abraham, 2003; Krajewska, 2004), polyacrylamide gel (Osiadacz et al. 1999), resin (Peralta-Zamora et al. 2003), porous glass beads (Rogalski et al. 1995), all the material used were listed by Duran et al. (2002)). The laccase immobilized on a mixture of dextran and Sepharose CL-6B gel remains active even in a medium containing only 3.5% of water (Milstein et al. 1989). To generate cofactors in situ, peroxidases were co-immobilized with glucose oxidase (Fig. 5) (Lee et al. 2003; Uyama et al. 1997). In addition, during the oxidation of thioanisole by the SBP, Van de Velde et al. (2000) reported that this system of co-immobilization enhances the enantioselectivity of the product from 24 to 50% (S). Another study has reported the co-immobilization of glucose oxidase and peroxidase on a biochip set in a channel with a continuous input of the reaction media. This system allows the online separation of the synthesized polymers (Lee et al. 2003).

In addition to traditionally used materials, peroxidases have been immobilized in microcapsules composed by nano-structured layers assembling enzyme and polyelectrolytes (LbL: Layer by Layer) (Ghan et al. 2003, 2004) (Fig. 4). This form of immobilization makes possible to have an important concentration of the

Table 6 Influence of phenol-substituent on phenolic compounds polymerization performances by Horseradish peroxidise

Molecule	Mw	Yield	Reference
Phenols	1,400	–	(Dordick et al. 1987)
p-methoxyphenol	2,000	–	
p-cresol	1,900	–	
p-chlorophenol	600	–	
2,6-dimethylphenol	500	–	
4,4'-biphenol	400	–	
Aniline	1,700	–	
1-naphtol	Very high	–	
2-naphtol	2,000	–	
p-tert-butylphenol	1,900	–	
p-phenylphenol	26,000	–	
4-cyclohexylphenol	1,200	86	(Mita et al. 2004)
4-t-butylphenol	1,674	98	
4-isopropylphenol	2,380	99	
4-methoxyphenol	1,159	100	
p-CH_3	6	7,920	(Kurioka et al. 1994)
p-n-C_3H_7	70	12,200	
p-i-C_3H_7	95	11,960	
p-t-C_4H_9	24	10,800	
p-n-C_5H_{11}	92	10,400	
p-n-C_7H_{15}	90	10,080	
o-i-C_3H_7	0	–	
m-i-C_3H_7	0	–	
H	72	12,2500	

Mw weight-average molecular mass

enzyme in small volume. Due to its selective permeability, the small molecules can easily diffuse from a compartment to another while the polymers are either trapped inside or adsorbed on the surface (Antipov et al. 2002).

Immobilization on supports or encapsulation may affects the affinity of the enzyme for its substrate. To remedy this problem, the technique of immobilization in a membrane reactor has been proposed and successfully applied to laccases (Durante et al. 2004).

4.2 Effect of Substrate

For the same enzyme, the type of substrate modifies the molecular weight and yield of obtained polymers (Table 6). Mita et al. (2003) observed that the molecular weight increases as a function of the log P of phenolic species. This may due to the difference in the polymer solubility in the solvent. These authors observed in the presence of HRP, that the percentage of the phenylens units (C–C) bridge increases compared to

Table 7 Influence of nature and redox potential of the substrate on phenolic compounds polymerization by Laccase of *Trametes versicolor*

Substrate	Substituent	Yield (%)	Redox potential (V)
Caffeic acid	–CH–CH–COOH	99	+ 0.53
Catechol	–	100	+0.53
2,6-hydroxybenzoic acid	–COOH	0	+1.0
3-hydroxybenzoic acid	–COOH	50	+1.02
3-Hydrocyphenylacetic aid	–CH$_2$–COOH	3	+1.06
4-hydroxybenzaldyde	–COH	0	–
Hydroxytyrosol	–CH$_2$–CH$_2$OH	88–100	–
Methylcathechol	–CH$_3$	98	+0.77
Protocatechuic acid	–COOH	46	+0.79
Synergic acid	–OCH$_3$	87	+0.68
	–COOH		
m-Tyrosol	–CH$_2$CH$_2$OH	39	–

Adapted from (Canfora et al. 2008) with kind permission of © The American Chemical Society (2008)

oxiphenylens units (C–O) when the hydrophobic parameters of the substrate (π) increase (Mita et al. 2004). The efficiency of the enzymatic transformation is strictly dependent on the chemical features of the phenolic substrate such as the number of OH groups, the nature and the molecular weight of the substituents and their position on the aromatic ring (*ortho* or *para*).

The low or absent reactivity of some phenols might be explained by their molecular structure and their higher redox potential (Table 7). In particular the substituents that behave as electron-donating (e.g –CH$_3$ or –CH$_2$CH$_3$) are those easier to give up an electron, to decrease oxidation potential and to increase oxidation rate. By contrast substituents that are electron—withdrawing (e.g. –NO$_2$, –Cl, –COOH) are harder to give up an electron, and usually they increase oxidation potential and decrease oxidation rate (Canfora et al. 2008).

To overcome the low solubility of substrates, some authors used charged polymers and micelles as vectors. This vectorization allows the increase of substrate solubility and avoid the parasite connections in polymers (Samuelson et al. 2003).

Kim et al. (2004b) observed that the use of water-soluble polymers like poly(ethylene glycol) (PEG), poly(methyl-2-oxazoline) (PMOZ) or poly(2-ethyl-2-oxazoline) (PEOZ), as a vectorization agents, favorizes the C–C coupling. For the HRP-catalyzed polymerization of phenol in water, the use of PEG as template increases the regioselectivity to 90% in favor of phenylene units. The regioselectivity enhancement can be due to an oriented alignment via hydrogen bonding interaction of phenolic OH groups and ether oxygen atoms of the PEG. The polymer was obtained in high yields as a precipitate complexed with PEG units (Kim et al. 2003a, b). A homogeneous reaction media and a high molecular weight Mw > 10^6 were obtained with Pluronic F68 (EG76-PG29- EG76) as a template (Kim et al. 2008).

Carbon nanotubes (CNTs) were recently used as template for the HRP-catalyzed polymerization of phenol in water. The polymerization reaction was conducted in the presence of the *p*-hydroquinone (HQ)- linked CNTs, and hence, polyphenol was grafted through the HQ moiety on the CNT surface. The phenol monomer was regioselectively polymerized to possess mainly the thermally stable oxyphenylene units. The high regioselectivity was considered to be the result of the shielding of ortho or para position of phenol from attack by adsorption onto the surface of CNTs (Yun Peng et al. 2009).

4.3 Effect of pH

The effect of pH is due to a compromise between two opposed processes (Xu 1997): in one hand, the increase of enzyme activity at high pH provokes an important difference between the potential of the substrate and that of the copper T1 of the laccase; in the other hand the decrease of the enzyme activity due to the fixation of hydroxyl anion on the copper sites T2 and T3. The effect of the pH on the polymerization reaction can be related to the nature of the substrate. For substrates having a redox potential independent of pH (ABTS, $K_4Fe(CN)_6$), laccase activity decreases with the increase of pH. However, for the other substrates the laccase activity depends almost on the laccase type (Xu 1997).

Fungal laccases typically exhibit pH optimal in the range from 3.5 to 5.0 when the substrates are hydrogen atom donor compound, and the pH-dependence curve is bell-shaped (Baldrian 2004; Koroleva et al. 2001; Kurniawati and Nicell 2008; Shleev et al. 2005). However, the optimum pH of the plant's laccases can be higher than 9 (Gianfreda et al. 1999). Those ranges of pH do not still correspond to the optimal conversion microenvironment during polymerization reaction. Indeed, Kim and Nicell (2006a) reported that, in the presence of laccase from *Trametes versicolor*, the best conversion yield (60%) is obtained at a pH of 5, while the optimal pH of this enzyme is from 6 to 8.

Moreover, pH influences the conversion rate and the molecular weight of obtained oligomers (Fig. 6) (Ikeda et al. 1998). Nevertheless this influence is variable according to the used reactional media (Anthoni et al. 2008; Dubey et al. 1998; Oguchi et al. 2000; Uyama et al. 1997).

Few data are available on the effect of pH on the structure of the synthesized polymers. Roubaty et al. (1978) reported that pH affects the type of linkage between monomeric units. An acid pH promotes the C–C linkage while C–O linkage is abundant at basic pH.

4.4 Effect of Ionic Species

According to Xu et al. (1998), small exogenous molecules like OH^-, F^- or Cl^- are able to be binding on the cupric site T_2 and to inhibit the electric transfer between site T_1 and the cluster T_2/T_3 of laccase. In addition, it has been demonstrated that molecule with chelating properties, like flavonoids, may inactive laccases. This inhibition can be avoided by the addition of the Cu^{2+} ions (Fenoll et al. 2003).

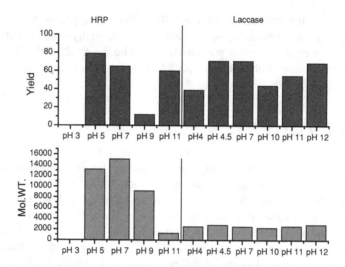

Fig. 6 Impact of pH variation on polymerization of phenol by Horseradish peroxidase (HRP) or laccase in 60% acetone or 50% acetone, respectively (Ikeda et al. 1998), Mol.WT.: molecular weight

The effect of inorganic salts and heavy metals on the polymerization of phenols, by HRP and LTV, was studied and discussed (Maalej-Kammoun et al. 2009; Wagner and Nicell 2002). The presence of NaCl, $CaCl_2$, $MgCl_2$, NH_4Cl and $(NH_4)_2SO_4$ decreases conversion rate compared to reactions realized in distilled water. Also, the presence of sulfide (H_2S) inhibits the oxidation of phenol by enzymes, while the presence of Mn(II) and Zn(II) promotes the activity of HRP (Kim and Nicell 2006b). Murugesan et al. (2009) reported that laccase activity was completely inhibited by Fe^{2+} at 0.5 mM.

4.5 Effect of Temperature

The optimal temperature for laccase-catalyzed polymerization typically ranges from 50 to 70°C (Baldrian 2006). There are few fungal laccases with an optimal oxidative profile below 35°C (Ko et al. 2001). Several studies approached the influence of the temperature on performance of the oxidoredutases catalyzed the enzymatic polymerization of phenols (Aktas 2005; Aktas et al. 2000; Aktas and Tanyolaç 2003; Anthoni et al. 2008; Guresir et al. 2005). Through these studies, it has been shown that temperature affects the activity of laccases, the conversion rate, the stability and the solubility of phenols. The intensity of this effect is variable according to the used substrate and the enzyme. As an example, Kim and Nicell (2006b) observed a better conversion of bisphenol A by LTV at 45°C then at 25°C while the optimal temperature of this laccase is 25°C. Using HRP and phenol as enzyme and substrate respectively, polymerization reaction was carried out at 10, 40, 50 and 60°C (Akita et al. 2001). High yields of polymerization (80%) are obtained as the temperature is inferior to 40°C. Beyond this temperature, the yield decreases (31% at 60°C). Such results could be explained by the decrease of the catalytic activity of HRP at 60°C and therefore due to the thermal deactivation of the apo-protein. In the same way, for rutin polymerization, a higher weight-average molecular mass index (I_M) and heterogeneity in polymer size were observed also at 10°C compared to those at 20 and 50°C. These results could be explained by a thermal deactivation of the laccase, which leads to a lower concentration of radical species. In such conditions, the elongation of the chains would be favored. The diminution of I_M at 50°C could be attributed to a larger concentration of radical species which favor coupling reactions and so, the termination of the synthesis. The comparison of these data to that reported by Kurisawa et al. (2003c) with the laccase from *Myceliophtora*, indicating that the effect of temperature is independent to the origin of laccase. An increase of temperature favors the synthesis of polymers. The highest production of polymers was reached at 50°C, when the solubility of rutin and the laccase activity are favored.

4.6 Effect of Solvent

Different works reported the effect of the solvent on the enzymatic polymerization of simple phenols (Ikeda et al. 1996a, b; Intra et al. 2005; Mita et al. 2003). These studies indicated that the nature and the proportion of the solvent modify the yield, the I_M and the structure of the polymers. For more complex phenols like flavonoids, there are only one work that discussed the effect of the nature and proportion of solvent (Anthoni et al. 2008).

In aqueous media, the polymerization reactions are limited by the weak solubility of phenolic compounds. In organic solvent, the activity of oxidoreductases is reduced due to their denaturation in such environment (Rodakiewicz-Nowak 2000). For that, the use of cosolvent was a good alternative to aqueous or organic solvents.

The ratio of organic solvent/buffer affects drastically the yield, the solubility and the average molecular weight of obtained polymers (Anthoni et al. 2008; Ikeda et al. 1996a, 1998; Kurisawa et al. 2003e; Mita et al. 2003) (Table 7). A simple variation of the percentage of used dioxane from 10 to 85% in water leads to an increase of the molecular weight from 500 to 26000 (Dordick et al. 1987). The enzymatic activity decreases with the increase of the dioxane percentage.

The addition of chloroform in the reaction media allows increasing of Mw of formed polymers. This behavior is due to interactions between the enzyme, the solvent, the monomer and the polymer, particularly to the decrease of the enzyme affinity to the substrate (Ikeda et al. 1996b).

To investigate the effect of the solvent on the enzymatic polymerization of rutin, Anthoni et al. (2008) tested different media (water, acetonitrile, dimethylformamide, methanol, ethanol, acetone, tetrahydrofurane). The highest and the

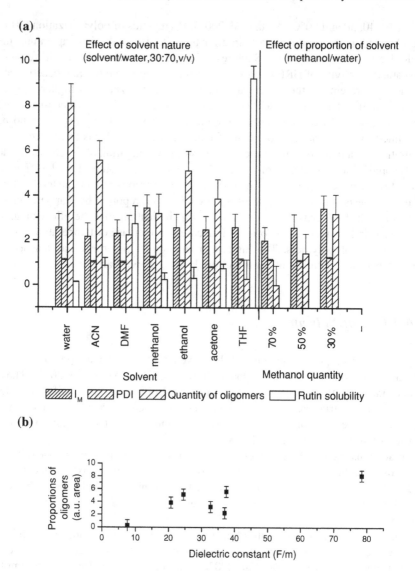

Fig. 7 Effect of nature and ratio of solvent on weight-average molecular mass index (I_M), polydispersity (PDI), production of oligomers and solubility of rutin (**a**), and influence of the dielectric constant of the cosolvent on the proportion of oligomers (**b**). Rutin (3 g/l) oligomerization was catalyzed by laccase from *Trametes versicolor* (3 U/ml), for 24 h, in solvent/water media (30:70, v/v), at 20°C. Reproduced from (Anthoni et al.) with kind permission of © Rasayan Journal of Chemistry (2008)

lowest production of polymers were obtained in water (8.11 a.u. area) and tetra-hydrofurane/water media (0.28 a.u. area). These results are correlated to the dielectric constant of the cosolvent. Higher the dielectric constant, lower the enzyme activity and polymer production (Fig. 7).

Fig. 8 Mechanism of substrate oxidation in presence of mediator

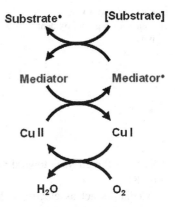

Ionic liquids are used as alternative to buffer or organic solvents in phenol polymerization reactions. The use of ionic liquids is motivated by two major advantages: maintain enzyme activity and stability (Tavares et al. 2008) and increase production of polymer and molecular weight (Eker et al. 2009).

The solvents used during the polymerization reaction also affects the structure of polymers and the selectivity of the reaction whether for HRP or laccase. In fact, when the log P of the solvent increases, the proportion of phenylene units decreases (Mita et al. 2004; Oguchi et al. 2000). This behavior is variable according to the nature of substrates. In the case of totarol oligomerization by LTV, C–C linkage is favored when the hydrophobicity of the cosolvent decreases. The effect of solvent on selectivity could be attributed to different interactions of the solvent molecules with the radical intermediates during their coupling reactions, without any real interference with the laccase oxidative action. If this had been the case, similar effect would have been observed using chemical oxidants. To determine the real origin of this selectivity, the enzymatic reaction was compared to a chemical once using FeCl$_3$ or MnO$_2$ as catalysts. The chemical reaction did not show any sterioselectivity, which suggests that the solvent affects the selectivity through its effect on the enzyme (Ncanana et al. 2007). The nature of the solvent effects have been discussed by other authors. Aromatic solvents like benzene and toluene lead to the formation of dimer structures (C1–C1' linkage) while solvent like *tert*-amyl alcohol and chloroform favored C1–C3' linkage (Intra et al. 2005).

4.7 Effect of Mediator

Oxidoreductases, especially laccases, possess relative low redox potential (≤ 0.8 V) whereas some phenolic substrates having higher redox potential cannot be oxidized directly by laccases (Chelikani et al. 2009). Nevertheless, this limitation has been overcome by using redox mediators. The presence of certain small-molecular weight compounds, which act as redox mediators expand the catalytic activity of the enzyme toward more recalcitrant compounds (Barreca et al. 2003) (Fig. 8).

Mediators act as electrons shuttles, providing the oxidation of complex substrates, which do not enter the active site due to the steric hindrances. Once oxidized by the enzyme and stabilized in more or less stable radicals, mediators diffuse far away from the enzymatic pocket and enable the oxidation of target compounds that in principle are not substrates of the enzyme because of their high size or high redox potential (Bourbonnais et al. 1997; Shingo Kawai and Umezawa 1989).

To enhance the redox reaction between the substrate and the enzyme, henothiazine-10-propionic acid was successfully used as mediator for the enzymatic polymerization of anacardic acid (Chelikani et al. 2009). In 2-propanol, the highest production yield (61%) of polymer (molecular weight = 3900 Da) was observed.

Nonylphenol (NP) and bisphenol A (BPA) were polymerized by laccase from *Coriolopsis polyzona* using 2,2'-azino-bis(3-ethylbenzthiazoline-6-sulfonic acid) (ABTS) as mediator. The polymerization of NP produced dimers, trimers, tetramers and pentamers which had molecular weights of 438, 656, 874 and 1092 amu, respectively. The polymerization of BPA produced dimers, trimers and tetramers which had molecular weights of 454, 680 and 906 amu (Cabana et al. 2007).

The presence of mediators affects also the architecture of the formed polymers. During the polymerization of 2,6-dimethylphenol and syringic acid by the laccase of *Trametes versicolor* in the presence and the absence of two mediators (2,2',6,6'-tetramethylpiperidine-N-oxyl radical (TEMPO) and 2,2'-azino-bis(3-ethylbenzthiazoline)-6-sulfonic acid (ABTS)), Marjasvaara et al. (2006) noticed that the synthesized oligomers had more variation in their structures than oligomers obtained without mediator.

4.8 Effect of Conduit Mode

The alimentation in substrate and especially in co-substrate, O_2 for laccases and H_2O_2 for peroxidises, may control the polymerization reaction. Generally, the O_2 is initially introduced into the reactional media. However, H_2O_2 is added during the reaction in two modes: controlled or not with a probe (Seelbach et al. 1997; van de Velde et al. 2001).

H_2O_2 is the indispensable co-substrate and in the same times an inhibitor of the peroxidasic activity at high concentrations. Its addition in the reaction medium has been the subject of several studies (Akita et al. 2001; Arnao et al. 1990; Nicell et al. 1995). To overcome the disadvantages of discontinuous alimentation, H_2O_2 was added in sufficient quantity at different intervals of the reaction (Akita et al. 2001). The addition of little concentrations, may promote the conversion yield. Control of hydrogen peroxide alimentation increases the conversion yield (Seelbach et al. 1997).

The majority of enzymatic polymerization reaction of phenolic compounds was carried on discontinuous mode (Table 3). To improve polymerization reaction performances, continuous mode was tested using two configurations. Pilaz et al. (2003) realized this reaction in an enzymatic-membrane reactor (EMR), where the addition of oxygen was assured by diffusion to overcome the air bubbling disadvantages. Seelbach et al. (1997) tested the EMR with a continuous elimination of produced polymers. Such procedure permitted to obtain a productivity of 119.5 g/(L.d) and a total turnover of 192.8 which is much better of the productivity (42.2 g/(L.d)) reached in batch mode. Batch mode on the other side allows higher turnover (three times higher).

Another procedure was experimented by Ayyagari et al. (1996, 1998) and consist on a continuous alimentation of the reactor with the substrate, the co-substrate and the enzyme. The reactor was coupled to an ultrafiltration module, which allows the separation of formed polymers. This process aims to control the molecular weight distribution, while minimizing the solvent effect and maximizing the enzyme utilization.

5 Effect of Enzymatic Polymerization on Phenolic Compounds Properties: Case of Flavonoids and Coumarins

Flavonoids are a class of phenolic secondary metabolites of plant that have recently received keen attention due to their antioxidant, antimicrobial and anti-carcinogenic properties. Many of these compounds are already used in pharmaceutical, cosmetic and food preparations. Unfortunately, the use of some of them is limited by their low solubility and stability in both lipophilic and aqueous media. Therefore, to improve these properties different techniques of derivatization are suggested, among them the enzymatic polymerization. As previously mentioned, this way allows the control of polymer structure due to the regioselectivity of the enzyme and can be conducted under mild operating conditions of temperature, pH and pressure. First data obtained in the case of the enzymatic polymerization of simple phenols and some flavonoids (rutin, catechin, quercetin, esculin) indicated that the polymerization of these compounds affects their radical scavenging activity (Anthoni et al. 2010, 2008; Desentis-Mendoza et al. 2006; Kurisawa et al. 2003a, d) and their solubility in the water (Anthoni et al. 2008, 2010; Bruno et al. 2005; Desentis-Mendoza et al. 2006; Kurisawa et al. 2003a, c, d). The magnitude of these effects depends on the degree of polymerization of the synthesized oligomers (Anthoni et al. 2008, 2010).

5.1 Antioxidant Activity

The interaction of flavonoids with many radicals has been described in several studies to determine the major elements of the antioxidant activity (Fig. 9).

Thanks to their low redox potential, flavonoids (Fl-OH) are thermodynamically capable of reducing free radical oxidants such as superoxide, peroxyl, alkoxyl and hydroxyl by hydrogen transfer.

Several modes of antioxidant activity action of flavonoids have been reported:

- Direct scavenging of free radicals: flavonoids are able to trap oxygen free radicals (X) by transfer of an electron or hydrogen

$$X^\bullet + AroH \rightarrow XH + Aro^\bullet$$
$$X^\bullet + AroH \rightarrow X^- + AroH^{\bullet+}$$

  The aryloxyl radical formed is stabilized by resonance. The unpaired electron can delocalize across the aromatic ring. But it can continue to evolve in several processes (dimerization, disproportionation, recombination with other radicals, reduction in parent molecule, quinone oxidation) or by reacting with radicals or other antioxidants or with biomolecules. The antiradical activity was correlated with the oxidation potential of flavonoids (van Acker et al. 1996).
- Chelating metals (Fe^{3+}, Cu^+)
  The antioxidant activity of flavonoids can be realized by the complexation of transition metals. They accelerate the formation of reactive oxygen species.
  Furthermore, the complexation of transition metals can improve the antioxidant activity of flavonoids by reducing their oxidation potential (Afanas et al. 2001).
- Enzyme inhibition
  Flavonoids are inhibitors of enzymes, including oxidoreductases that are involved in their catalytic cycle radical species (lipoxygenase, cyclooxygenase, monooxygenase, xanthine oxidase, protein kinase...).
  Enzymatic polymerization of phenolic compounds affects their biological properties. These properties, including antioxidant activities, may be dependent on the molecular weight of the synthesized polymers and the type and the position of the linkages ($\overline{M_w}$, PDI, C–C or C–O bridges). Moreover, depending on the used method for determining antiradical activity (AAPH, DPPH, ...) of polyphenols, results are controversial.

As an example, for rutin polymerization by laccase from *Pycnoporus coccineus*, *Pycnoporus sanguineus or Myceliophthora*, synthesized polymers presented a better inhibition of AAPH radical, compared to its monomer (Kurisawa et al. 2003b). However, Anthoni et al. (2008) reported that polyrutin, obtained through laccase from *Trametes versicolor* polymerization, have a weaker DPPH radical scavenging activity compared to rutin. For other phenolic compounds, like catechin, kaempferol, esculin and 8-hydroquinoline, polymerization enhances inhibition effects against free radical including oxidation of low-density lipoprotein (LDL), DPPH radical (Desentis-Mendoza et al. 2006).

Using xanthine oxidase inhibition test, it was well established that enzymatic polymerization of phenolic compounds (rutin, esculin, catechin) may increase antioxidant activity (Anthoni et al. 2008, 2010; Kurisawa et al. 2003a).

5.2 Solubility

As it was mentioned previously, one of the problem uses of phenolic compounds, is their weak solubility. The first result of enzymatic polymerization reported that the obtained polymers of rutin and esculin are 4200-folds and 189-folds more water soluble than rutin and esculin, respectively (Anthoni et al. 2008, 2010).

To explain the high solubility of rutin oligomers, the interactions and especially the H-bonds between these molecules and water were evaluated by a molecular modeling study (Anthoni et al. 2008). In water, rutin showed a folded structure where the rhamnose and the ring A became closer together. Six intermolecular H-bonds were established with the water molecules. The hexamer revealed an unfolded structure where sugars offer a large contact with the surrounding solvent. The number of intermolecular H-bonds between the hexamer and water molecules was evaluated to 94. This dense network of H-bonds could explain the high solubility of the hexamer compared to rutin.

6 Conclusion

The use of laccases and peroxidases for the phenolic compounds polymerization provides an interesting alternative, to synthesize molecules with different molecular weight and properties. However, those reactions are affected by many factors (temperature, pH, solvent, the enzyme origin, the substrate structure, the polymerization mode …). Those factors often have antagonistic effects, such as the effect of organic solvent that on the one side favors the substrates solubility, but in the other side alters the enzymatic activity. The optimization of those reactions needs a polydisciplinary approach. Taking into account the major interest of phenolic compounds polymers, the coming years will see an intensification of investigations in this area, which will lead to the emerging of new processes.

Acknowledgments The authors would like to acknowledge Julie Anthoni and Ghada Ben Rhouma for their contribution in the realization of this book. Also, we would like to acknowledge Rasayan Journal for their permission to reuse figures.

References

Abou El Hassan MAI, Touw DJ, Wilhelm AJ et al (2000) Stability of monoHER in an aqueous formulation for i.v. administration. Int J Pharm 211:51–56

Afanas, rsquo, eva IB et al (2001) Enhancement of antioxidant and anti-inflammatory activities of bioflavonoid rutin by complexation with transition metals. Biochem Pharmacol 61:677–684

Akita M, Tsutsumi D, Kobayashi M et al (2001) Structural change and catalytic activity of horseradish peroxidase in oxidative polymerization of phenol. Biosci Biotechnol Biochem 65:1581–1588

Akkara JA, Kaplan DL, Samuelson et al (1992) Method for synthesizing an enzyme-catalyzed polymerized monolayer. vol 5.143.828. Lowell, Massachusetts: The United States of America as represented by the secretary of the army

Aktas N (2005) Optimization of biopolymerization rate by response surface methodology (RSM). Enzyme Microb Technol 37:441–447

Aktas N, Kibarer G, Tanyolaç A (2000) Effects of reaction conditions on laccase-catalyzed alpha-naphthol polymerization. J Chem Technol Biotechnol 75:840–846

Aktas N, Tanyolaç A (2003) Reaction conditions for laccase catalyzed polymerization of catechol. Bioresour Technol 87:209–214

Angerer PS, Studer A, Witholt B et al (2005) Oxidative polymerization of a substituted phenol with ion-paired horseradish peroxidase in an organic solvent. Macromol 38:6248–6250

Anthoni J, Humeau C, Maia ER et al (2010) Enzymatic synthesis of oligoesculin: Structure and biological activities characterizations. Eur Food Res Technol 231:571–579

Anthoni J, Lionneton F, Wieruszeski JM et al (2008) Investigation of enzymatic oligomerization of rutin. Rasayan J Chem 1:718–731

Antipov AA, Sukhorukov GB, Leporatti S et al (2002) Polyelectrolyte multilayer capsule permeability control. Colloids Surf A 198–200:535–541

Arnao MB, Acosta M, del Rio JA et al (1990) A kinetic study on the suicide inactivation of peroxidase by hydrogen peroxide. Biochim Biophys Acta, Protein Struct Mol Enzymol 1041:43–47

Ayyagari M, Akkara JA, Kaplan DL (1996) Characterization of phenolic polymers synthesized by enzyme-mediated reactions in bulk solvents and at oil-water interfaces. Mater Sci Eng, C 4:169–173

Ayyagari M, Akkara JA, Kaplan DL (1998) Solvent-enzyme-polymer interactions in the molecular-weight control of poly(m-cresol) synthesis in nonaqueous media. In: Gross RA, Kaplan DL, Swift G (ed) Enzymes in polymer synthesis. ACS symposium series 684, Washington

Ayyagari MS, Marx KA, Tripathy SK et al (1995) Controlled free-radical polymerization of phenol derivatives by enzyme-catalyzes reactions in organic solvents. Macromol 28:5192–5197

Baldrian P (2004) Purification and characterization of laccase from the white-rot fungus Daedalea quercina and decolorization of synthetic dyes by the enzyme. Appl Microbiol Biotechnol 63:560–563

Baldrian P (2006) Fungal laccases-occurrence and properties. FEMS Microbiol Rev 30:215–242

Barreca AM, Fabbrini M, Galli C et al (2003) Laccase/mediated oxidation of a lignin model for improved delignification procedures. J Mol Catal B: Enzym 26:105–110

Bilici A, Kaya I, YildIrIm M et al (2010) Enzymatic polymerization of hydroxy-functionalized carbazole monomer. J Mol Catal B: Enzym 64:89–95

Bindhu LV, Abraham ET (2003) Immobilization of horseradish peroxidase on chitosan for use in nonaqueous media. J Appl Polym Sci 88:1456–1464

Bindhu LV, Emilia Abraham T (2003) Preparation and kinetic studies of surfactant-horseradish peroxidase ion paired complex in organic media. Biochem Eng J 15:47–57

Bourbonnais R, Paice MG, Freiermuth B et al (1997) Reactivities of various mediators and laccases with kraft pulp and lignin model compounds. Appl Environ Microbiol 63:4627–4632

Bruno FF, Akkara JA, Samuelson LA et al (1995) Enzymatic mediated synthesis of conjugated polymers at the Langmuir trough air-water interface. Langmuir 11:889–892

Bruno FF, Nagarajan S, Nagarajan R et al (2005) Biocatalytic synthesis of water-soluble oligo(catechins). J Macromol Sci Pure Appl Chem 42:1547–1554

Burton SG (2003) Laccases and phenol oxidase in organic synthesis—A review. Curr Org Chem 7:1317–1331

Cabana H, Jiwan J-LH, Rozenberg R et al (2007) Elimination of endocrine disrupting chemicals nonylphenol and bisphenol A and personal care product ingredient triclosan using enzyme preparation from the white rot fungus Coriolopsis polyzona. Chemosphere 67:770–778

Canfora L, Iamarino G, Rao MA et al (2008) Oxidative Transformation of Natural and Synthetic Phenolic Mixtures by Trametes versicolor Laccase. J Agric Food Chem 56:1398–1407

Carrea G (1984) Biocatalysis in water-organic solvent two-phase systems. Trends Biotechnol 2:102–106

Chelikani R, Kim YH, Yoon DY et al (2009) Enzymatic polymerization of natural anacardic acid and antibiofouling effects of polyanacardic acid coatings. Appl Biochem Biotechnol 157:263–277

Claus H (2004) Laccases: Structure, reactions, distribution. Micron 35:93–96

Dec J, Bollag JM (1990) Detoxification of substituted phenols by oxidoreductive enzymes through polymerization reactions. Arch Environ Contam Toxicol 19:543–550

Desentis-Mendoza RM, Hernandez-Sanchez H, Moreno A et al (2006) Enzymatic polymerization of phenolic compounds using laccase and tyrosinase from Ustilago maydis. Biomacromolecules 7:1845–1854

Dordick JS, Marletta MA, Klibanov AM (1987) Polymerization of phenols catalyzed by peroxidase in nonaqueous media. Biotechnol Bioeng 30:31–36

Dubey S, Singh D, Misra RA (1998) Enzymatic synthesis and various properties of poly(catechol). Enzyme Microb Technol 23:432–437

Duran N, Rosa MA, D'Annibale A et al (2002) Applications of laccases and tyrosinases (phenoloxidases) immobilized on different supports: a review. Enzyme Microb Technol 31:907–931

Durante D, Casadio R, Martelli L et al (2004) Isothermal and non-isothermal bioreactors in the detoxification of waste waters polluted by aromatic compounds by means of immobilised laccase from Rhus vernicifera. J Mol Catal B: Enzym 27:191–206

Eker B, Zagorevski D, Zhu G et al (2009) Enzymatic polymerization of phenols in room-temperature ionic liquids. J Mol Catal B: Enzym 59:177–184

Faure A, Bouillant ML, Jacoud C et al (1995) Phenolic derivatives related to lignin metabolism as substrates for azospirillum laccase activity. Phytochemistry 42:357–359

Fenoll LG, Garcia-Ruiz PA, Varon R et al (2003) Kinetic Study of the Oxidation of Quercetin by Mushroom Tyrosinase. J Agric Food Chem 51:7781–7787

Friedman M, Jurgens HS (2000) Effect of pH on the Stability of Plant Phenolic Compounds. J Agric Food Chem 48:2101–2110

Fulcrand H, Doco T, Es-Safi N-E et al (1996) Study of the acetaldehyde induced polymerisation of flavan-3-ols by liquid chromatography-ion spray mass spectrometry. J Chromatogr A 752:85–91

Ghan R, Shutava T, Patel A, et al (2003) Layer-by-layer engineered microreactors for bio-polymerization of 4-(2-aminoethyl) phenol hydrochloride. In: Materials research society symposium—proceedings 782:241

Ghan R, Shutava T, Patel A et al (2004) Enzyme-catalyzed polymerization of phenols within polyelectrolyte microcapsules. Macromol 37:4519–4524

Gianfreda L, Xu F, Bollag J-M (1999) Laccases: a useful group of oxidoreductive enzymes. Biorem J 3:1–25

Guresir M, Aktas N, Tanyolac A (2005) Influence of reaction conditions on the rate of enzymic polymerization of pyrogallol using laccase. Process Biochem 40:1175–1182

Havsteen BH (2002) The biochemistry and medical significance of the flavonoids. Pharmacol Ther 96:67–202

Heim KE, Tagliaferro AR, Bobilya DJ (2002) Flavonoid antioxidants: chemistry, metabolism and structure-activity relationships. J Nutr Biochem 13:572–584

Hinckley G, Mozhaev VV, Budde C et al (2002) Oxidative enzymes possess catalytic activity in systems with ionic liquids. Biotechnol Lett 24:2083–2087

Hudson EP, Eppler RK, Clark DS (2005) Biocatalysis in semi-aqueous and nearly anhydrous conditions. Curr Opin Biotechnol 16:637–643

Ikeda R, Sughihara J, Uyama H et al (1998) Enzymatic oxidative polymerization of 4-hydroxybenzoic acid derivatives to poly(phenylene oxide)s. Polym Int 47:295–301

Ikeda R, Sugihara J, Uyama H et al (1996a) Enzymatic oxidative polymerization of 2,6-dimethylphenol. Macromol 29:8702–8705

Ikeda R, Tsujimoto T, Tanaka H et al (2000) Man-made urushi. Proc Jpn Acad 76:155–160

Ikeda R, Uyama H, Kobayashi S (1996b) Novel synthetic pathway to a poly(phenylene oxide). laccase-catalyzed oxidative polymerization of syringic acid. Macromol 29:3053–3054

Intra A, Nicotra S, Riva S, et al (2005) Significant and unexpected solvent influence on the selectivity of laccase-catalyzed coupling of tetrahydro-2-naphtol derivatives. Adv Synth Catal, 347:973–977

Job D, Dunford HB (1976) Substituent effect on the oxidation of phenols and aromatic amines by horseradish peroxidase compound I. Eur J Biochem 66:607–614

Kamiya N, Inoue M, Goto M et al (2000) Catalytic and structural properties of surfactant-horseradish peroxidase complex in organic media. Biotechnol Progr 16:52–58

Khmelnitsky YL, Gladilin AK, Roubailo VL et al (1992) Reversed micelles of polymeric surfactants in nonpolar organic solvents. A new microheterogeneous medium for enzymatic reactions. Eur J Biochem 206:737–745

Kim S, Silva C, Zille A et al (2009) Characterisation of enzymatically oxidised lignosulfonates and their application on lignocellulosic fabrics. Polym Int 58:863–868

Kim Y-J, Chung JE, Kurisawa M et al (2004a) New Tyrosinase Inhibitors, (+)-Catechin-Aldehyde Polycondensates. Biomacromolecules 5:474–479

Kim Y-J, Nicell, JA (2006a) Impact of reaction conditions on the laccase-catalyzed conversion of bisphenol A. Bioresour Technol 97:1431–1442

Kim Y-J Uyama H, Kobayashi S (2003a) Regioselective synthesis of poly(phenylene) as a complex with poly(ethylene glycol) by template polymerization of Phenol in water. Macromol 36:5058–5060

Kim YH, An ES, Song BK et al (2003b) Polymerization of cardanol using soybean peroxidase and its potential application as anti-biofilm coating material. Biotechnol Lett 25:1521–1524

Kim YJ, Nicell JA (2006b) Impact of reaction conditions on the laccase-catalyzed conversion of bisphenol A. Bioresour Technol 97:1431–1442

Kim YJ, Shibata K, Uyama H et al (2008) Synthesis of ultrahigh molecular weight phenolic polymers by enzymatic polymerization in the presence of amphiphilic triblock copolymer in water. Polymer 49:4791–4795

Kim YJ, Uyama H, Kobayash S (2004b) Enzymatic template polymerization of phenol in the presence of water-soluble polymers in an aqueous medium. Polym J 36:992–998

Ko KEM, Leem Leem YE et al (2001) Purification and characterization of laccase isozymes from the white-rot basidiomycete Ganoderma lucidum. Appl Microbiol Biotechnol 57:98–102

Kobayashi S, Higashimura H (2003) Oxidative polymerization of phenols revisited. Prog Polym Sci 28:1015–1048

Kobayashi S, Kurioka H, Uyama H (1996) Enzymatic synthesis of a soluble polyphenol derivative from 4,4'-biphenyldiol. Macromol Rapid Comm 17:503–508

Kobayashi S, Uyama H, Kimura S (2001) Enzymatic polymerization. Chem Rev 101:3793–3818

Koroleva OV, Yavmetdinov IS, Shleev SV et al (2001) Isolation and study of some properties of laccase from the Basidiomycetes Cerrena maxima. Biochemistry (Moscow) 66:618–622

Krajewska B (2004) Application of chitin- and chitosan-based materials for enzyme immobilizations: a review. Enzyme Microb Technol 35:126–139

Kurioka H, Komatsu I, Uyama H et al (1994) Enzymatic oxidative polymerization of alkylphenols. Macromol Rapid Comm 15:507–510

Kurisawa M, Chung JE, Kim YJ et al (2003a) Amplification of antioxidant activity and xanthine oxidase inhibition of catechin by enzymatic polymerization. Biomacromolecules 4:469–471

Kurisawa M, Chung JE, Uyama H et al (2003b) Enzymatic synthesis and antioxidant properties of poly(rutin). Biomacromolecules 4:1394–1399

Kurisawa M, Chung JE, Uyama H et al (2003c) Enzymatic synthesis and antioxidant properties of poly(rutin). Biomacromolecules 4:1394–1399

Kurisawa M, Chung JE, Uyama H et al (2003d) Laccase-catalyzed synthesis and antioxidant property of poly(catechin). Macromol Biosci 3:758–764

Kurisawa M, Chung JE, Uyama H et al (2003e) Laccase-catalyzed synthesis and antioxidant property of poly(catechin). Macromol Biosci 3:758–764

Kurniawati S, Nicell JA (2008) Characterization of Trametes versicolor laccase for the transformation of aqueous phenol. Bioresour Technol 99:7825–7834

Lalot T, Brigodiot M, Maréchal E (1999) A kinetic approach to acrylamide radical polymerization by horse radish peroxidase-mediated initiation. Polym Int 48:288–292

Laszlo JA, Compton DL (2002) Comparison of peroxidase activities of hemin, cytochrome c and microperoxidase-11 in molecular solvents and imidazolium-based ionic liquids. J Mol Catal B: Enzym 18:109

Lee MY, Srinivasan A, Ku B et al (2003) Multienzyme catalysis in microfluidic biochips. Biotechnol Bioeng 83:20–28

Ma H-L, Kermasha S, Gao J-M et al (2009) Laccase-catalyzed oxidation of phenolic compounds in organic media. J Mol Catal B Enzym 57:89–95

Maalej-Kammoun M, Zouari-Mechichi H, Belbahri L et al (2009) Malachite green decolourization and detoxification by the laccase from a newly isolated strain of Trametes sp. Int Biodeterior Biodegrad 63:600–606

Mai C, Majcherczyk A, Hüttermann A (2000) Chemo-enzymatic synthesis and characterization of graft copolymers from lignin and acrylic compounds. Enzyme Microb Technol 27:167–175

Marjasvaara A, Torvinen M, Kinnunen H et al (2006) Laccase-catalyzed polymerization of two phenolic compounds studied by matrix-assisted laser desorption/ionization time-of-flight and electrospray ionization fourier transform ion cyclotron resonance mass spectrometry with collision-induced dissociation experiments. Biomacromolecules 7:1604–1609

Martinek K, Levashov AV, Khmelnitsky NL et al (1982) Colloidal solution of water in organic solvents: a microheterogeneous medium for enzymatic reactions. Science 218:889–891

Michizoe J, Goto M (2001) S. F. Catalytic activity of lactase hosted in reversed micelles. J Biosci Bioeng 92:67–71

Milstein O, Nicklas B, Hüttermann A (1989) Oxidation of aromatic compounds in organic solvents with laccase from Trametes versicolor. Appl Microbiol Biotechnol 31:70–74

Mita N, Tawaki SI, Uyama H et al (2002) Enzymatic oxidative polymerization of phenol in an aqueous solution in the presence of a catalytic amount of cyclodextrin. Macromol Chem Phys 203:127–130

Mita N, Tawaki SI, Uyama H et al (2003) Laccase-catalyzed oxidative polymerization of phenols. Macromol Biosci 3:253–257

Mita N, Tawaki SI, Uyama H et al (2004) Precise structure control of enzymatically synthesized polyphenols. Bull Chem Soc Jpn 77:1523–1527

Morozova OV, Shumakovich GP, Gorbacheva MA, Shleev SV, Yaropolov AI (2007) "Blue" laccases. Biochemistry (Moscow) 72:1136–1150

Murugesan K, Kim Y-M, Jeon J-R et al (2009) Effect of metal ions on reactive dye decolorization by laccase from Ganoderma lucidum. J Hazard Mater 168:523–529

Mustafa R, Muniglia L, Rovel B et al (2005) Phenolic colorants obtained by enzymatic synthesis using a fungal laccase in a hydro-organic biphasic system. Food Res Int 38:995–1000

Nabid MR, Zamiraei Z, Sedghi R et al (2010) Synthesis and characterization of poly(catechol) catalyzed by porphyrin and enzyme. Polym Bull 64:855–865

Nayak PL (1998) Enzyme-catalyzed polymerization: an opportunity for innovation. Des Monomers Polym 1:259–284

Ncanana S, Baratto L, Roncaglia L et al (2007) Laccase-mediated oxidation of totarol. Adv Synth Catal 349:1507–1513

Nicell JA, Saadi KW, Buchanan ID (1995) Phenol polymerization and precipitation by horseradish peroxidase enzyme and an additive. Bioresour Technol 54:302–310

Nugroho Prasetyo E, Kudanga T, Østergaard L et al (2010) Polymerization of lignosulfonates by the laccase-HBT (1-hydroxybenzotriazole) system improves dispersibility. Bioresour Technol 101:5054–5062

Oguchi T, Tawaki SI, Uyama H et al (2000) Enzymatic synthesis of soluble polyphenol. Bull Chem Soc Jpn 73:1389–1396

Osiadacz J, Al-Adhami AJH, Bajraszewska D et al (1999) On the use of *Trametes versicolor* laccase for the conversion of 4-methyl-3-hydroxyanthranilic acid to actinocin chromophore. J Biotechnol 72:141–149

Paradkar VM, Dordick JS (1994a) Aqueous-like activity of alpha-chymotrypsin dissolved in nearly anhydrous organic solvents. J Am Chem Soc 116:5009–5010

Paradkar VM, Dordick JS (1994b) Mechanism of extraction of chymotrypsin into isooctane at very low concentrations of aerosol OT in the absence of reversed micelles. Biotechnol Bioeng 43:529–540

Peralta-Zamora P, Pereira CM, Tiburtius ERL et al (2003) Decolorization of reactive dyes by immobilized laccase. Appl Catal B, 42:131

Pilz RP, Hammer EH, Schauer FS et al (2003) Laccase-catalysed synthesis of coupling products of phenolic substrates in different reactors. Appl Microbiol Biotechnol 60:708–712

Poulos TL (1993) Peroxidases. Curr Opin Biotechnol 4:484–489

Premachandran RS, Banerjee S, Wu XK et al (1996) Enzymatic synthesis of fluorescent naphtol-based polymers. Macromol 29:6452–6460

Rao AM, John VT, Gonzales RD et al (1993) Catalytic and interfacial aspects of enzymatic polymer synthesis in reversed micellar systems. Biotechnol Bioeng 41:531–540

Reihmann H, Ritter (2001) Oxidative Copolymerisation of para-functionalized phenols catalyzed by Horseradish peroxidase and Thermocrosslinking via Diels-Alder and (1 + 3) Cycloaddition. Macromol Biosci 1:85–90

Reihmann MH, Ritter H (2000) Enzymatically catalyzed synthesis of photocrosslinkable oligophenols. MacromolChemPhys 201:1593–1597

Rittstieg K, Suurnakki A, Suortti T et al (2002) Investigations on the laccase-catalyzed polymerization of lignin model compounds using size-exclusion HPLC. Enzyme Microb Technol 31:403–410

Riva S (2006) Laccases: blue enzymes for green chemistry. Trends Biotechnol 24:219–226

Rodakiewicz-Nowak J (2000) Phenols oxidizing enzymes in water-restricted media. Top Catal 11–12:419–434

Rogalski J, Jozwik E, Hatakka A et al (1995) Immobilization of laccase from *Phlebia radiata* on controlled porosity glass. J Mol Catal A Chem 95:99–108

Roubaty JL, Bréant M, Lavergne M et al (1978) Mechanism of the oxidative coupling of phenols: influence of pH upon the electrochemical oxidation of xylenol in methanol. Makromol Chem 179:1151–1157

Samuelson LA, Tripathy SK, Bruno F, et al (2003) Enzymatic polymerization of anilines or phenols around a template. vol US 6.569.651 B1. The United States of America as represented by the secretary of the army, lowell, Massachusetts

Seelbach K, Van Deurzen MPJ, Van Rantwijk F et al (1997) Improvement of the total turnover number and space-time yield for chloroperoxidase catalyzed oxidation. Biotechnol Bioeng 55:283–288

Shingo K, Tashiaki U, Higushi T (1989) Oxidation of methoxylated benzyl alcohols by laccase of Coriolus versicolor in the presence of syringaldehyde. Wood Res 76:10–16

Shleev S, Jarosz-Wilkolazka A, Khalunina A et al (2005) Direct electron transfer reactions of laccases from different origins on carbon electrodes. Bioelectrochem 67:115–124

Shleev S, Reimann CT, Serezhenkov V et al (2006) Autoreduction and aggregation of fungal laccase in solution phase: possible correlation with a resting form of laccase. Biochimie 88:1275–1285

Solomon EIAAJ, Yoon J (2008) O_2 reduction to H_2O by the multicopper oxidases. Dalton Transactions 3921–3932

Solomon EI, Sundaram UM, Machonkin TE (1996) Multicopper oxidases and oxygenases. Chem Rev 96:2563–2605

Tanaka T, Takahashi M, Hagino H et al (2010) Enzymatic oxidative polymerization of methoxyphenols. Chem Eng Sci 65:569–573

Tavares APM, Rodriguez O, Macedo EA (2008) Ionic liquids as alternative co-solvents for laccase: study of enzyme activity and stability. Biotechnol Bioeng 101:201–207

Tonami H, Uyama H, Kobayashi S et al (2000) Chemoselective oxidative polymerization of m-ethynylphenol by peroxidase catalyst to a new reactive polyphenol. Biomacromolecules 1:149–151

Uyama H, Kurioka H, Kobayashi S (1997) Novel bienzymatic catalysis system for oxidative polymerization of phenols. Polym J 29:190–192

Uyama H, Maruichi N, Tonami H et al (2002) Peroxidase-catalyzed oxidative polymerization of bisphenols. Biomacromolecules 3:187–193

Uzan E, Portet B, Lubrano C et al (2011) Pycnoporus laccase-mediated bioconversion of rutin to oligomers suitable for biotechnology applications. Appl Microbiol Biotechnol 90:97–105

van Acker SABE, de Groot MJ, van den Berg D-J et al (1996) A quantum chemical explanation of the antioxidant activity of flavonoids. Chem Res Toxicol 9:1305–1312

Van De Velde F, Lourenco ND, Bakker M et al (2000) Improved operational stability of peroxidases by coimmobilization with glucose oxidase. Biotechnol Bioeng 69:286–291

van de Velde F, van Rantwijk F, Sheldon RA (2001) Improving the catalytic performance of peroxidases in organic synthesis. Trends Biotechnol 19:73–79

Van Deurzen MPJ, Van Rantwijk F, Sheldon RA (1997) Selective oxidations catalyzed by peroxidases. Tetrahedron 53:13183–13220

Van Rantwijk F, Lau RM, Sheldon RA (2003) Biocatalytic transformations in ionic liquids. Trends Biotechnol 21:131–138

Wagner M, Nicell JA (2002) Impact of dissolved wastewater constituents on peroxidase-catalyzed treatment of phenol. J Chem Technol Biotechnol 77:419–428

Welinder KG (1992) Superfamily of plant, fungal and bacterial peroxidases. Curr Opin Struct Biol 2:388–393

Witayakran S, Ragauskas AJ (2009) Synthetic applications of laccase in green chemistry. Adv Synth Catal 351:1187–1209

Xia Z, Yoshida T, Funaoka M (2003) Enzymatic synthesis of polyphenols from highly phenolic lignin-based polymers (lignophenols). Biotechnol Lett 25:9–12

Xu F (1996) Oxidation of phenols, anilines, and benzenethiols by fungal laccases: correlation between activity and redox potentials as well as halide inhibition. Biochemistry 35:7608–7614

Xu F (1997) Effects of redox potential and hydroxide inhibition on the pH activity profile of fungal laccases. J Biol Chem 272:924–928

Xu F, Berka RM, Wahleithner JA et al (1998) Site-directed mutations in fungal laccase: effect on redox potential, activity and pH profile. Biochem J 334(Pt 1):63–70

Yang Z, Pan W (2005) Ionic liquids: green solvents for nonaqueous biocatalysis. Enzyme Microb Technol 37:19–28

Yoshida T, Lu R, Han S et al (2009) Laccase-catalyzed polymerization of lignocatechol and affinity on proteins of resulting polymers. J Polym Sci, Part A: Polym Chem 47:824–832

Yu JH, Klibanov AM (2006) Co-lyophilization with D-proline greatly enhances peroxidase's stereoselectivity in a non-aqueous medium. Biotechnol Lett 28:555

Yun Peng HL, Zhang Xingyuan, Li Yuesheng, Liu Shiyong (2009) CNT templated regioselective enzymatic polymerization of phenol in water and modification of surface of MWNT thereby. J Polym Sci, Part A: Polym Chem 47:1627–1635

Zaragoza-Gasca P, Gimeno M, Hernández JM et al (2011) Novel photoconductive polyfluorophenol synthesized by an enzyme. J Mol Catal B Enzym 72:25–27

Zhu QY, Holt RR, Lazarus SA et al (2002) Stability of the Flavan-3-ols epicatechin and catechin and related dimeric procyanidins derived from cocoa. J Agric Food Chem 50:1700–1705

Index

M. Ghoul and L. Chebil, *Enzymatic Polymerization of Phenolic Compounds by Oxidoreductases*, SpringerBriefs in Green Chemistry for Sustainability, DOI: 10.1007/978-94-007-3919-2, © The Author(s) 2012